QC 929.24 D75 2005

Drought and Water Crises

Science, Technology, and Management Issues

Drought and Water Crises

Science, Technology, and Management Issues

Edited by
Donald A. Wilhite

Boca Raton London New York Singapore

A CRC title, part of the Taylor & Francis imprint, a member of the Taylor & Francis Group, the academic division of T&F Informa plc.

Published in 2005 by
CRC Press
Taylor & Francis Group
6000 Broken Sound Parkway NW, Suite 300
Boca Raton, FL 33487-2742

© 2005 by Taylor & Francis Group
CRC Press is an imprint of Taylor & Francis Group

No claim to original U.S. Government works
Printed in the United States of America on acid-free paper
10 9 8 7 6 5 4 3 2 1

International Standard Book Number-10: 0-8247-2771-1 (Hardcover)
International Standard Book Number-13: 978-0-8247-2771-0 (Hardcover)
Library of Congress Card Number 2004061861

This book contains information obtained from authentic and highly regarded sources. Reprinted material is quoted with permission, and sources are indicated. A wide variety of references are listed. Reasonable efforts have been made to publish reliable data and information, but the author and the publisher cannot assume responsibility for the validity of all materials or for the consequences of their use.

No part of this book may be reprinted, reproduced, transmitted, or utilized in any form by any electronic, mechanical, or other means, now known or hereafter invented, including photocopying, microfilming, and recording, or in any information storage or retrieval system, without written permission from the publishers.

For permission to photocopy or use material electronically from this work, please access www.copyright.com (http://www.copyright.com/) or contact the Copyright Clearance Center, Inc. (CCC) 222 Rosewood Drive, Danvers, MA 01923, 978-750-8400. CCC is a not-for-profit organization that provides licenses and registration for a variety of users. For organizations that have been granted a photocopy license by the CCC, a separate system of payment has been arranged.

Trademark Notice: Product or corporate names may be trademarks or registered trademarks, and are used only for identification and explanation without intent to infringe.

Library of Congress Cataloging-in-Publication Data

Drought and water crisis : science, technology, and management issues / Donald A. Wilhite.
p. cm. — (Books in soils, plants, and the environment ; v. 86)
1. Droughts. 2. Water-supply—Risk assessment. I. Wilhite, Donald A. II Series.
ISBN 0-847-2771-1 (alk. paper)

QC929.24.D75 2005
363.34'9297—dc22

2004061861

Taylor & Francis Group
is the Academic Division of T&F Informa plc.

Visit the Taylor & Francis Web site at
http://www.taylorandfrancis.com

and the CRC Press Web site at
http://www.crcpress.com

To Myra, Addison, Shannon, Suzanne,
Benjamin, and my grandson, Gabriel

Contents

Editor's Preface

When I began my professional career at the University of Nebraska–Lincoln in 1979, I intended to direct my research and outreach program at the emerging field of climate impact science. It was fortuitous that a large portion of the United States, including the Great Plains, Upper Midwest, and Pacific Northwest, had recently come out of an intense but somewhat short-lived drought during 1976-1977. This drought spawned a research-oriented workshop held at the University of Nebraska in 1979 that focused on drought impacts and the development of agricultural drought strategies for that area and similar regions. I was given the opportunity to work with the project team to design the workshop content and develop pre-workshop materials. Although I had focused my graduate studies on climate variability and the climatology of drought, my intent was for drought to be only one of several climate-related subject areas I would address in my career. The workshop led to two follow-up drought projects directed at an evaluation of governmental drought response policies.

Twenty-five years later, I am still researching and writing about drought. There must be something fascinating about this subject to capture my imagination for the past quarter century. As I became more engaged in the subject, both as a climate scientist and a geographer, I became more and more intrigued by its complexity and the challenges of detecting, responding to, and preparing for this "natural" hazard. Why was drought such a poorly understood concept? What was the role of the science community in addressing this issue? Why were governments so poorly prepared for drought? Why were governmental policies for dealing with drought nonexistent? From both a scientific and a policy perspective, we have made considerable progress in addressing many of the issues associated with improving how society manages drought. Much remains to be done, however; especially with drought's interconnections to issues of integrated water management, sustainable development, climate change, water scarcity, environmental degradation, transboundary water conflicts, population growth, and poverty, to name just a few.

Drought and Water Crises: Science, Technology, and Management Issues is an attempt to explain the complexities of drought and the role of science, technology, and management in resolving many of the perplexing issues associated with drought management and the world's expanding water crises. Tremendous advances have been made in the past decade in our ability to monitor and detect drought and communicate this information to decision makers at all levels. Why are decision makers not fully using this information for risk mitigation? Better planning and mitigation tools are also available today to help governments and other groups develop drought mitigation plans. How can we make these methodologies more readily available and adaptable? In the agricultural and urban sectors, new water-conserving technologies are being applied that allow more efficient use of water. How can we promote more widespread adoption of these technologies and their use during non-drought periods? Progress is being made on improving the reliability of seasonal drought forecasts to better serve decision makers in the management of water and other natural resources. How can these seasonal forecasts be made more reliable and expressed in ways to better meet the needs of end users? These and other questions are addressed by the contributors to this volume. The information herein will better equip the reader with the knowledge necessary to take action to reduce societal vulnerability to drought.

In the past, most regions possessed a buffer in their water supply so periods of drought were not necessarily associated with water shortages, although impacts were often quite severe. The crisis management approach to drought management, although ineffective in reducing societal vulnerability, allowed societies to muddle through to the next drought episode. That buffer no longer exists for most locations. Water shortages are widespread in both developing and developed countries and in more humid as well as arid climates—even in years with relatively normal precipitation. Drought only serves to exacerbate these water shortages and conflicts between users. Droughts of lesser magnitude are also resulting in greater impacts—a clear sign that more people and sectors are at greater risk today than in the past. When societies are faced with a long-term drought, such as has been occurring in the western United States over the past 6 years, governments are desperate to identify longer term solutions. Unfortunately, this interest often quickly wanes when precipitation returns to normal—a return to the "hydro-illogical" mentality.

All drought-prone nations should adopt a more risk-based, proactive policy for drought management. To make progress, we must first recognize that drought has both a natural and a social dimension. Second, we must involve natural, biological, and social scientists in the formulation and implementation of drought preparedness plans and policies. This book collates considerable information from diverse disciplines with the goal of furthering drought preparedness planning and reducing societal vulnerability to drought.

Contributors

Marianne Alden is a researcher with the Adaptation and Impacts Group, Meteorological Service of Canada, Environment Canada, in Waterloo, Ontario. Her research interests include surface water management and policy, climate change impacts on water quality and quantity, and phenology.

Linda Botterill is a postdoctoral fellow at the National Europe Centre at the Australian National University in Canberra. Her research interest is agricultural policy in Australia and the European Union, with a focus on the policy development process in developed economies. She has a particular interest in drought policy and rural adjustment.

Margie Buchanan-Smith has worked for many years in the humanitarian aid sector. Her experience ranges from policy research to operational management, from drought and natural disasters to war and violent conflict. She was a research fellow at the Overseas Development Institute in London and at the Institute of Development Studies at the University of Sussex. She was also head of ActionAid's Emergencies Unit between 1995 and 1998. She now works freelance.

David W. Cash is the Director of Air Policy in the Massachusetts Executive Office of Environmental Affairs. Before this position associate at the John F. Kennedy School of Government at Harvard University in Cambridge, Massachusetts, USA, and a Lecturer in Environmental Science and Public Policy. He received a Ph.D. in Public Policy at Harvard with his dissertatio and post-graduate research focusing on water management in the U.S. Great Plains.

Luiz F. N. Cavalcanti received his master's degree in city and regional planning from Georgia Tech and degrees in civil and environmental engineering from Federal University of Minas Gerais, Brazil. His interests focus on drought management and preparedness. He helped to develop the indicators and triggers for Georgia's first drought plan and conducted a nationwide evaluation of U.S. state drought plans.

Francis H. S. Chiew is an associate professor in environmental engineering in the Department of Civil and Environmental Engineering at the University of Melbourne in Victoria, Australia. Dr. Chiew has more than 15 years experience in research, teaching, and consulting in hydrology and water resources and related disciplines. He is currently a program leader (climate variability) in the Cooperative Research Centre for Catchment Hydrology. His interests include hydroclimatology, hydrological modeling, and urban stormwater quality.

Stewart J. Cohen is a scientist with the Adaptation and Impacts Research Group of the Meteorological Service of Canada in Environment Canada and an adjunct professor with the Institute for Resources, Environment and Sustainability, University of British Columbia. He has more than 20 years research experience in climate change impacts and adaptation and has organized case studies throughout Canada. He has contributed to the Intergovernmental Panel on Climate Change (IPCC) and served as an adviser and lecturer for impacts and adaptation research and training programs in China, Europe, and the United States, as well as the United Nations Environment Programme.

Michael J. Coughlan is head of the National Climate Centre in the Australian Bureau of Meteorology. He has worked on several national and international programs dealing with drought and other aspects of climate variability and change; he has also occupied positions within the U.S. National Oceanographic and Atmospheric Administration, the World Climate Research Programme, and the World Meteorological Organization.

Susan Cuddy is a researcher within the Integrated Catchment Assessment and Management (iCAM) Centre at The Australian National University, Canberra, Australia, and in the Integrated Catchment Management directorate at CSIRO Land and Water, Canberra, Australia. She has been involved in the development and design of environmental software to support natural resource management for more than 20 years. Her main research interests are in knowledge representation and the "packaging" of science for a range of audiences.

Randall M. Dole is the director of the NOAA Climate Diagnostics Center in Boulder, Colorado, USA. His research interests include extended-range weather and climate predictions, applications of climate information and forecasts, and explaining causes for drought and other extreme climate events. He has made numerous presentations on drought causes, characteristics, and predictions, and is interagency co-lead for the "Climate Variability and Change" element of the U.S. Climate Change Science Program.

Robert A. Halliday is a consulting engineer in Saskatoon, Saskatchewan, and a former director of Canada's National Hydrology Research Centre. His interests concern interjurisdictional water management, floodplain management, and effects of climate on water resources. He has served on International Joint Commission boards and other Canada–U.S. water-related entities and has worked on water management projects in many countries.

Michael J. Hayes is a climate impacts specialist with the National Drought Mitigation Center and a research associate professor in the School of Natural Resources at the University of Nebraska, Lincoln, USA. His work focuses on strategies to reduce drought risk through improved drought monitoring, planning, and identification of appropriate drought mitigation activities.

Katharine L. Jacobs is an associate professor in the Soil, Water and Environmental Science Department at the University of Arizona in Tucson, USA, and deputy director of SAHRA, the Center for Sustainability of Semi-Arid Region Hydrology and Riparian Areas. Her research areas include climate and water management, water policy, and use of science in decision making. She formerly was director of the Tucson office of the Arizona Department of Water Resources.

Tony Jakeman is a professor in the Centre for Resource and Environmental Studies and director of the Integrated Catchment Assessment and Management Centre of The Australian National University, Canberra. He has been an environmental modeler for 28 years and has more than 300 publications in the open literature. His current research interests include integrated assessment methods for water and associated land resource problems, as well as modeling of water supply and quality problems, including in ungauged catchments.

Ke Li Dan is a professor, senior engineer, and former director of the Department of Water Resources Administration of MWR of China. He was the organizer and chairman of the drafting committee of the Water Law of China and is the president of the Water Law Association of China, a member of IWRA, and an executive member of AIDA. He has been engaged in water administration and water resources management for more than 20 years.

Cody L. Knutson is a water resources specialist with the National Drought Mitigation Center, located in the School of Natural Resources at the University of Nebraska, Lincoln, USA. His work incorporates both physical and social sciences to foster better understanding of drought vulnerability and management.

Grace Koshida is a researcher with the Adaptation & Impacts Research Group (Environment Canada) in Toronto, Canada. Her research activities focus on drought impacts and drought adaptations, high-impact weather events, and climate change impacts on water resources.

Douglas Le Comte is a meteorologist and drought specialist with NOAA's Climate Prediction Center in Camp Springs, Maryland, USA. His work focuses on drought monitoring and forecasting. He spearheaded development of the U.S. Drought Monitor in 1999 and played an active role in the development of the U.S. Seasonal Drought Outlook, for which he is the lead forecaster.

Rebecca Letcher is a research fellow at the Integrated Catchment Assessment and Management Centre at The Australian National University in Canberra. Her research activities have focused on the application and development of integrated assessment methods for water resource management, particularly participatory model building approaches.

Abdel Maarouf is a biometeorologist with the Adaptation & Impacts Research Group (Environment Canada) in Toronto. He conducts collaborative research on environmental stresses on human health, such as extremes of heat and cold, increased risk of infectious diseases due to climate change, and impacts of weather disasters on urban health.

Manuel Menéndez Prieto is the scientific and technical coordinator at CEDEX (Experimental Center on Public Works, Spanish Ministry of the Environment). He is a lecturer in the Polytechnic University of Madrid. His research has focused on hydrological extreme events. Currently, he is in charge of technical coordination of the Spanish contribution to the implementation strategy of the European Union's Water Framework Directive.

Karl Monnik worked at the ARC-Institute for Soil, Climate and Water in South Africa, where he was responsible for agrometeorological research. He was involved in a number of national drought policy committees and organized and participated in several national and international drought meetings. He recently moved to the Bureau of Meteorology in Australia, where he is involved in meteorological observation networks.

Linda D. Mortsch is a senior researcher with the Adaptation and Impacts Research Group of Environment Canada, located in Ontario at the University of Waterloo in the Faculty of Environmental Studies. Her research interests include climate impact and adaptation assessment in water resources and wetlands. She has been an active participant in the Intergovernmental Panel on Climate Change process and has published numerous reports and papers on climate variability and change.

Blair E. Nancarrow is the director of the Australian Research Centre for Water in Society in CSIRO Land and Water in Western Australia. She specializes in social investigations and public involvement programs in water resources management and community input to policy making. She is particularly interested in the development of processes to incorporate social justice in environmental decision making.

Neville Nicholls leads the Climate Forecasting Group at the Bureau of Meteorology Research Centre in Melbourne, Australia. Since 1972 he has been researching the nature, causes, impacts, and predictability of climate variations and change, especially for the Australian region.

Theib Y. Oweis is a project manager and senior irrigation and water resources management scientist of the International Center for Agricultural Research in the Dry Areas (ICARDA), Aleppo, Syria. He manages, conducts research, and runs capacity building programs on managing water resources in agriculture under scarcity and drought in the dry areas—mainly Central Asia, West Asia, and North Africa. His research focuses on supplemental irrigation, water harvesting, and improving water productivity, and his activities involve collaboration with national, regional, and international organizations.

Phil Pasteris is a supervisory physical scientist with the USDA's National Water and Climate Center in Portland, Oregon,

USA. He is responsible for the production and distribution of water supply forecasts for the western United States and management of the agency's climate program.

Colin Polsky is an assistant professor in the Graduate School of Geography and the George Perkins Marsh Institute at Clark University in Worcester, Massachusetts, USA. Dr. Polsky was educated at the University of Texas at Austin, Pennsylvania State University, and Harvard University. He blends quantitative and qualitative methods to study social vulnerability to the effects of climate change.

Roger S. Pulwarty is a research scientist at the National Oceanic and Atmospheric Administration Climate Diagnostics Center at the University of Colorado in Boulder, USA. His research and practical interests are in assessing the role of climate and weather in society–environment interactions and in designing effective local, national, and international services to address associated risks. From 1998 to 2002 he directed the NOAA/Regional Integrated Sciences and Assessments (RISA) Program.

Kelly T. Redmond is the deputy director and regional climatologist of the Western Regional Climate Center at the Desert Research Institute in Reno, Nevada, USA. He earned a B.S. degree in physics from the Massachusetts Institute of Technology and M.S. and Ph.D. degrees in meteorology from the University of Wisconsin in Madison. His research and professional interests span every facet of climate and climate behavior, climate's physical causes and behavior, how climate interacts with other human and natural processes, and how such information is acquired, used, communicated, and perceived.

Anne C. Steinemann is a professor of civil and environmental engineering and director of the Center for Water and Watershed Studies at the University of Washington in Seattle, USA. She was formerly associate professor at Georgia Tech and visiting scholar at the Scripps Institution of Oceanography. Her areas of expertise include drought indicators and triggers, drought plans, and climate forecasts for water management.

Mark Svoboda is a climatologist with the National Drought Mitigation Center and a research scientist in the School of Natural Resources at the University of Nebraska in Lincoln, USA. His responsibilities include providing expertise on climate and water management issues by working with state and federal agencies, international governments, the media, and the private sector. He also maintains the NDMC's drought monitoring activities. Mark serves as one of the principal authors of both the U.S. Drought Monitor and the North American Drought Monitor.

Amy Vickers is an engineer, water conservation consultant, public policy advisor, and author of the *Handbook of Water Use and Conservation: Homes, Landscapes, Businesses, Industries, Farms* (WaterPlow Press). She is president of Amy Vickers & Associates, Inc., in Amherst, Massachusetts, USA. She holds an M.S. in engineering from Dartmouth College and a B.A. in philosophy from New York University.

Donald A. Wilhite is founder and director of the National Drought Mitigation Center and a professor in the School of Natural Resources at the University of Nebraska, Lincoln, USA. His research and outreach activities are centered on issues of drought monitoring, mitigation, planning, and policy, and he has collaborated with numerous countries and regional and international organizations on matters related to drought management.

Virginia Wittrock is a climatologist/research scientist with the Saskatchewan Research Council in Saskatoon, Saskatchewan, Canada. Her research interests are in the areas of descriptive climatology (e.g., research into the drought situation in Saskatchewan and the Canadian prairies), climate change research as it pertains to impacts and adaptation strategies, and teleconnection patterns. She has served as a member of the board of directors in the Saskatchewan Provincial Branch of the Canadian Water Resources Association.

Zhang Hai Lun is the former deputy director of Nanjing Research Institute of Hydrology and Water Resources, Ministry of Water Resources, and former chief of the Natural Resources Division of UNESCAP. He has long been involved in research activities in the field of water resources assessment and planning, hydrological analysis, and strategy on flood control and water management.

Zhang Shi Fa is a retired professor and adviser of the Department of Hydrology and Water Resources of Nanjing Hydraulic Research Institute. His research fields are focused on statistics analysis, water resource assessment and planning, drought analysis, and mitigation, including the study of historical drought in China.

Acknowledgments

Drought and Water Crises: Science, Technology, and Management Issues is the result of the efforts of many persons who have been working diligently over the past 2 years to bring this volume to fruition. The book was conceived through discussions between me and Susan Lee of Marcel Dekker, Inc. Susan was a pleasure to work with during manuscript development and most responsive to my myriad questions. My interactions with Matt Lamoreaux and others at CRC Press were extremely positive and helpful throughout the latter stages of this project.

I would especially like to thank the contributors to this volume. These colleagues were carefully chosen for their expertise, the quality of their research throughout their professional careers, and the contribution of their research efforts and experiences to the theme of this book. I appreciate their responsiveness to the deadlines I imposed in the preparation of the initial draft of their chapters and their receptivity to suggested edits and modifications.

Finally, I would like to thank Deb Wood and Ann Fiedler of the National Drought Mitigation Center for their many contributions to the preparation of the manuscript. I have valued Deb's editing skills throughout my tenure at the University of Nebraska. This book is just one of many manuscripts to which Deb has contributed her many talents and skills over the years. Ann's organizational skills are unsurpassed and have facilitated the book preparation process. She was also responsible for the final formatting of the manuscript for CRC Press. Their flexibility and sense of humor throughout this process have been most appreciated.

Part I

Overview

1

Drought as Hazard: Understanding the Natural and Social Context

DONALD A. WILHITE AND
MARGIE BUCHANAN-SMITH

CONTENTS

I. INTRODUCTION

Drought is an insidious natural hazard that results from a deficiency of precipitation from expected or "normal" that, when extended over a season or longer, is insufficient to meet the demands of human activities and the environment. Drought by itself is not a disaster. Whether it becomes a disaster depends on its impact on local people and the environment. Therefore, the key to understanding drought is to understand both its natural and social dimensions.

Drought is a normal part of climate, rather than a departure from normal climate (Glantz, 2003). The latter view of drought has often led policy and other decision makers to treat this complex phenomenon as a rare and random event. This perception has typically resulted in little effort being targeted toward those individuals, population groups, economic sectors, regions, and ecosystems most at risk (Wilhite, 2000). Improved drought policies and preparedness plans that are proactive rather than reactive and that aim at reducing risk rather than responding to crisis are more cost-effective and can lead to more sustainable resource management and reduced interventions by government (Wilhite et al., 2000a; see also Chapter 5).

The primary purpose of this chapter is to discuss drought in terms of both its natural characteristics and its human dimensions. This overview of the concepts, characteristics, and impact of drought will provide readers with a foundation for a more complete understanding of this complex hazard and how it affects people and society and, conversely, how societal use and misuse of natural resources and government policies can exacerbate vulnerability to this natural hazard. In other words, we are promoting a holistic and multidisci-

plinary approach to drought. This discussion is critical to an understanding of the material presented in the science and technology section of this volume (Part II) as well as in the various case studies presented in Part III.

We use the term *hazard* to describe the natural phenomenon of drought and the term *disaster* to describe its negative human and environmental impacts.

II. DROUGHT AS HAZARD: CONCEPTS, DEFINITION, AND TYPES

Drought differs from other natural hazards in several ways. First, drought is a slow-onset natural hazard, often referred to as a creeping phenomenon (Gillette, 1950). Because of the creeping nature of drought, its effects accumulate slowly over a substantial period of time. Therefore, the onset and end of drought are difficult to determine, and scientists and policy makers often disagree on the bases (i.e., criteria) for declaring an end to drought. Tannehill (1947) notes:

> We may truthfully say that we scarcely know a drought when we see one. We welcome the first clear day after a rainy spell. Rainless days continue for some time and we are pleased to have a long spell of fine weather. It keeps on and we are a little worried. A few days more and we are really in trouble. The first rainless day in a spell of fine weather contributes as much to the drought as the last, but no one knows how serious it will be until the last dry day is gone and the rains have come again … we are not sure about it until the crops have withered and died.

Should drought's end be signaled by a return to normal precipitation and, if so, over what period of time does normal or above-normal precipitation need to be sustained for the drought to be declared officially over? Do precipitation deficits that emerged during the drought event need to be erased for the event to end? Do reservoirs and groundwater levels need to return to normal or average conditions? Impacts linger for a considerable time following the return of normal precipitation; so is the end of drought signaled by meteorological or climatological factors, or by the diminishing negative human impact?

Second, the absence of a precise and universally accepted definition of drought adds to the confusion about whether a drought exists and, if it does, its degree of severity. Realistically, definitions of drought must be region and application (or impact) specific. Definitions must be region specific because each climate regime has distinctive climate characteristics (i.e., the characteristics of drought differ significantly between regions such as the North American Great Plains, Australia, southern Africa, western Europe, and northwestern India). Definitions need to be application specific because drought, like beauty, is largely defined by the beholder and how it may affect his or her activity or enterprise. Thus, drought means something different for a water manager, an agriculturalist, a hydroelectric power plant operator, and a wildlife biologist. Even within sectors there are many different perspectives of drought because impacts may differ markedly. For example, the impacts of drought on crop yield may differ greatly for maize, wheat, soybeans, and sorghum because each is planted at a different time during the growing season and has different sensitivities to water and temperature stress at various growth stages. This is one explanation for the scores of definitions that exist. For this reason, the search for a universal definition of drought is a rather pointless endeavor. Policy makers are often frustrated by disagreements among scientists on whether a drought exists and its degree of severity. Usually, policy makers' principal interest is the impact on people and the economy and the types of response measures that should be employed to assist the victims of drought.

Third, drought impacts are nonstructural and spread over a larger geographical area than are damages that result from other natural hazards such as floods, tropical storms, and earthquakes. This, combined with drought's creeping nature, makes it particularly challenging to quantify the impact, and may make it more challenging to provide disaster relief than for other natural hazards. These characteristics of drought have hindered development of accurate, reliable, and timely estimates of severity and impacts (i.e., drought early warning systems) and, ultimately, the formulation of drought preparedness plans. Similarly, emergency managers, who have the assignment of responding to drought, struggle to deal with the large spatial coverage usually associated with drought.

Drought is a temporary aberration, unlike aridity, which is a permanent feature of the climate. Seasonal aridity (i.e., a well-defined dry season) also must be distinguished from drought. Considerable confusion exists among scientists and policy makers on the differentiation of these terms. For example, Pessoa (1987) presented a map illustrating the frequency of drought in northeastern Brazil in his discussion of the impacts of and governmental response to drought. For a significant portion of the northeast region, he indicated that drought occurred 81–100% of the time. Much of this region is arid, and drought is an inevitable feature of its climate. However, drought is a temporary feature of the climate, so it cannot, by definition, occur 100% of the time.

Nevertheless, it is important to identify trends over time and whether drought is becoming a more frequent and severe event. Concern exists that the threat of global warming may increase the frequency and severity of extreme climate events in the future (IPCC, 2001). As pressure on finite water supplies and other limited natural resources continues to build, more frequent and severe droughts are cause for concern in both water-short and water-surplus regions where conflicts within and between countries are growing. Reducing the impacts of future drought events is paramount as part of a sustainable development strategy, a theme developed later in this chapter and throughout this volume.

Drought must be considered a relative, rather than absolute, condition. It occurs in both high- and low-rainfall areas and in virtually all climate regimes. Our experience suggests scientists, policy makers, and the public often associate drought only with arid, semiarid, and subhumid regions. In reality, drought occurs in most nations, in both dry and humid regions, and often on a yearly basis. The intensity, epicenter, and size of the area affected by drought will vary annually (see Chapter 12), but its presence is nearly always being felt. This reality supports the need for a national strategy (see Chapters 5 and 6).

A. Types of Drought

All types of drought originate from a deficiency of precipitation (Wilhite and Glantz, 1985). When this deficiency spans an extended period of time (i.e., *meteorological drought*), its existence is defined initially in terms of these natural characteristics. The natural event results from persistent large-scale disruptions in the global circulation pattern of the atmosphere (see Chapter 2). Exposure to drought varies spatially, and there is little, if anything, we can do to alter drought occurrence. However, the other common drought types (i.e., agricultural, hydrological, and socioeconomic) place greater

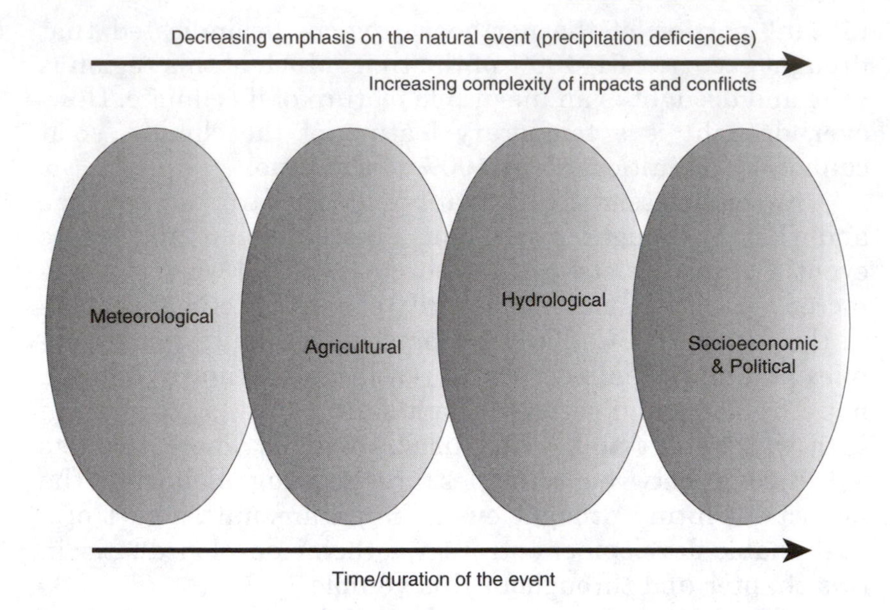

Figure 1 Natural and social dimensions of drought. (*Source*: National Drought Mitigation Center, University of Nebraska, Lincoln, Nebraska, USA.)

emphasis on human or social aspects of drought, highlighting the interaction or interplay between the natural characteristics of the event and the human activities that depend on precipitation to provide adequate water supplies to meet societal and environmental demands (see Figure 1). For example, *agricultural drought* is defined more commonly by the availability of soil water to support crop and forage growth than by the departure of normal precipitation over some specified period of time. No direct relationship exists between precipitation and infiltration of precipitation into the soil. Infiltration rates vary according to antecedent moisture conditions, slope, soil type, and the intensity of the precipitation event. Soils also vary in their characteristics, with some soils having a high water-holding capacity and others a low water-holding capacity. Soils with a low water-holding capacity are more drought prone.

Hydrological drought is even further removed from the precipitation deficiency because it is normally defined in terms of the departure of surface and subsurface water supplies from some average condition at various points in time. Like agricultural drought, no direct relationship exists between precipitation amounts and the status of surface and subsurface water supplies in lakes, reservoirs, aquifers, and streams because these components of the hydrological system are used for multiple and competing purposes (e.g., irrigation, recreation, tourism, flood control, hydroelectric power production, domestic water supply, protection of endangered species, and environmental and ecosystem preservation). There is also considerable time lag between departures of precipitation and when these deficiencies become evident in these components of the hydrologic system. Recovery of these components is also slow because of long recharge periods for surface and subsurface water supplies. In areas where the primary source of water is snowpack, such as in the western United States, the determination of drought severity is further complicated by infrastructures, institutional arrangements, and legal constraints. For example, reservoirs increase this region's resilience to drought because of the potential for storing large amounts of water as a buffer during dry years. However, the operating plans for these reservoirs try to accommodate the multiple uses of the water (e.g., protection of fisheries, hydroelectric power production, recreation and tourism, irrigation) and the priorities set by the U.S. Congress when the funds were allocated to construct the reservoir. The allocation of water between these various users is generally fixed and inflexible, making it difficult to manage a drought period. Also, legal agreements between political jurisdictions (i.e., states, countries) concerning the amount of water to be delivered from one jurisdiction to another impose legal requirements on water managers to maintain flows at certain levels. During drought, conflicts heighten because of limited available water. These shortages may result from poor water and land management practices that exacerbate the problem (e.g., see Chapters 10 and 12).

Socioeconomic drought differs markedly from the other types because it associates human activity with elements of meteorological, agricultural, and hydrological drought. This may result from factors affecting the supply of or demand for some commodity or economic good (e.g., water, grazing, hydroelectric power) that is dependent on precipitation. It may also result from the differential impact of drought on different groups within the population, depending on their access or entitlement to particular resources, such as land, and/or their access or entitlement to relief resources. Drought may fuel conflict between different groups as they compete for limited resources. A classic example in Africa is the tension, which may become violent in drought years, between nomadic pastoralists in search of grazing and settled agriculturalists wishing to use the same land for cultivation. The concept of socioeconomic drought is of primary concern to policy makers.

The interplay between drought and human activities raises a serious question with regard to attempts to define it in a meaningful way. It was previously stated that drought results from a deficiency of precipitation from expected or "normal" that is extended over a season or longer period of time and is insufficient to meet the demands of human activities and the environment. Conceptually, this definition assumes that the demands of human activities are in balance or harmony with the availability of water supplies during periods of normal or mean precipitation. If development demands exceed the supply of water available, demand may exceed supply even in years of normal precipitation. This can result in human-induced drought. In this situation, development can be sustained only through mining of groundwater and/or the transfer of water into the region from other watersheds. Is this practice sustainable in the long term? Should this situation be defined as "drought" or unsustainable development?

Drought severity can be aggravated by other climatic factors (such as high temperatures, high winds, and low relative humidity) that are often associated with its occurrence in many regions of the world. Drought also relates to the timing (i.e., principal season of occurrence, delays in the start of the rainy season, occurrence of rains in relation to principal crop growth stages) and effectiveness of the rains (i.e., rainfall intensity, number of rainfall events). Thus, each drought event is unique in its climatic characteristics, spatial extent, and impacts (i.e., no two droughts are identical). The area affected by drought is rarely static during the course of the event. As drought emerges and intensifies, its core area or epicenter shifts and its spatial extent expands and contracts. A comprehensive drought early warning system is critical for tracking these changes in spatial coverage and severity, as explained below.

B. Characterizing Drought and Its Severity

In technical terms, droughts differ from one another in three essential characteristics: intensity, duration, and spatial coverage. Intensity refers to the degree of the precipitation shortfall and/or the severity of impacts associated with the shortfall. It is generally measured by the departure of some climatic parameter (e.g., precipitation), indicator (e.g., reservoir levels), or index (e.g., Standardized Precipitation Index) from normal and is closely linked to duration in the determination of impact. These tools for monitoring drought are discussed in Chapter 3. Another distinguishing feature of drought is its duration. Droughts usually require a minimum of 2 to 3 months to become established but then can continue for months or years. The magnitude of drought impacts is closely related to the timing of the onset of the precipitation shortage, its intensity, and the duration of the event.

Droughts also differ in terms of their spatial characteristics. The areas affected by severe drought evolve gradually, and regions of maximum intensity (i.e., epicenter) shift from season to season. In larger countries, such as Brazil, China, India, the United States, or Australia, drought rarely, if ever, affects the entire country. During the severe drought of the 1930s in the United States, for example, the area affected by severe and extreme drought reached 65% of the country in 1934. This is the maximum spatial extent of drought in the period from 1895 to 2003. The climatic diversity and size of countries such as the United States suggest that drought is likely to occur somewhere in the country each year. On average 14% of the country is affected by severe to extreme drought annually. From a planning perspective, the spatial characteristics of drought have serious implications. Nations should determine the probability that drought may simultaneously affect all or several major crop-producing regions or river basins within their borders and develop contingencies for such an event. Likewise, it is important for governments to calculate the chances of a regional drought simultaneously affecting agricultural productivity and water supplies in their country and adjacent or nearby nations on which they depend for food supplies. A drought mitigation strategy that relies on the importation of food from neighboring countries may not be viable if a regional-scale drought occurs.

III. DROUGHT AS DISASTER: THE SOCIAL/POLITICAL CONTEXT

Drought, like all natural hazards, has both a natural and social dimension. The risk associated with drought for any region is a product of both the region's exposure to the event (i.e., probability of occurrence at various severity levels) and the vulnerability of society to the event. Vulnerability can be defined as "defenselessness, insecurity, exposure to risk, shocks and stress," and difficulty in coping with them (Chambers, 1989). It is determined by both micro- and macro-level factors, and it is cross-sectoral—dependent on economic, social, cultural, and political factors. Blaikie et al.'s (1994) disaster pressure model represents well the interaction of hazard with vulnerability (Figure 2). They explore vulnerability in terms of three levels. First, there are the root causes. These may be quite remote and are likely to relate to the underlying political and economic systems and structures. Second are the dynamic pressures, which translate the effects of the root causes into particular forms of insecurity. These pressures might include rapid population growth, rapid urbanization, and epidemics. As a result, unsafe conditions are created; for instance, through people living in dangerous locations and/or the state failing to provide adequate protection.

Understanding people's vulnerability to drought is complex yet essential for designing drought preparedness, mitigation, and relief policies and programs. At the micro level, determinants of vulnerability include:

- The physical asset base of the household—for example, land, livestock, cash
- Human capital—for example, productive labor
- Social capital—for example, claims that can be made on other households within the community, perhaps for productive resources, food, or labor

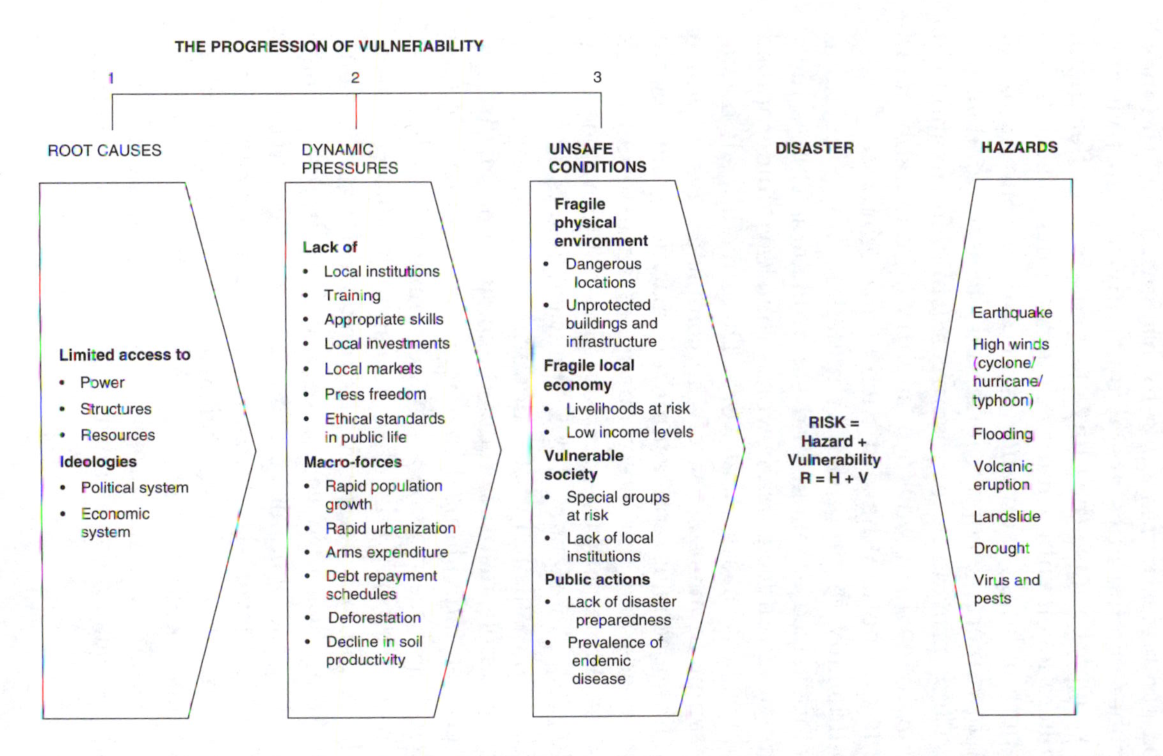

Figure 2 "Pressures" that result in disasters: the progression of vulnerability. (*Source*: *At Risk: Natural Hazards, People's Vulnerability, and Disasters*, P. Blaikie, T. Cannon, I. Davis, and B. Wisner, London: Routledge, 1994, Figure 2.1, p. 23.)

All of these refer to the asset base of the household. Generally speaking, the stronger and more diverse the household's asset base, the more drought resilient it is likely to be, and the greater its options in terms of switching between different livelihood strategies in response to drought. Thus, the most impoverished communities are also usually the most vulnerable to drought, because they have few assets to buffer them. At the macro level, vulnerability determinants include security, strength of local governance structures, accountability of the state to vulnerable populations, and the associated ability of the state to provide relief resources. Thus, for example, a population living in a war-torn country is inevitably more vulnerable to a natural hazard such as drought.

Traditionally, the approach to understanding vulnerability has emphasized economic and social factors. This is most evident in the livelihoods frameworks that have underpinned much vulnerability assessment work. These livelihoods frameworks attempt to make sense of the complex ways in which individuals, households, and communities achieve and sustain their livelihoods and the likely impact of an external shock such as drought on both lives *and* livelihoods[1] (Save the Children [UK], 2000; Young et al., 2001). Political factors and power relationships have usually been underplayed in these frameworks. For example, institutionalized exploitation and discrimination between individuals, households, and groups are often overlooked. Yet these may be a key determinant of whether a particular ethnic group will have access to productive assets such as land and to relief resources provided by government. Similarly, many war-torn countries are also drought prone. Understanding the dynamics and impact of the conflict—from national to local level—is critical to understanding the population's vulnerability to drought, as described in the case study of South Sudan presented later in Section V.B.

Some recent work has proposed how the political dimension of livelihoods analysis can be strengthened by including

---

[1] See, for example, www.livelihoods.org.

political economy analysis, explicitly including issues of power (Collinson, 2003). The international aid community's recent interest in a rights-based approach to development has the potential to strengthen further the political aspect of vulnerability analysis in developing countries. In a rights-based approach, one asks questions about the claims individuals or households are entitled to, identifies those responsible for meeting these claims—the "duty-bearers"—and is concerned with the persistent denial or violation of these rights, which renders an individual or group particularly vulnerable (O'Neill, 2003). For example, in Gujarat State in India there is institutionalized discrimination against the *dalits* (the schedule caste) and *kolis* (a tribal community). Not only are these groups denied access to some social infrastructure, but this institutionalized discrimination can quickly turn a relief program from progressive to regressive, with the poorest and most marginalized groups receiving the least assistance. In an emergency response, exploitative social structures and power relations simply reproduce, with even more devastating consequences as limited relief and rehabilitation resources are captured by the better off. A rights-based approach should reveal these underlying patterns of discrimination, and hence vulnerability, and may require some positive and controversial steps to be taken to challenge the status quo and prioritize the needs of these marginalized groups (see Buchanan-Smith, 2003a).

Understanding and measuring the vulnerability to drought of a population or of particular groups within that population is not an easy task. It requires an in-depth knowledge of the society and the relationships within that society. It is not a job for the newcomer. Instead, it benefits from long-term familiarity, yet the ability to remain objective. Also, vulnerability is not a static concept. Hence, no two droughts will have the same human impact. Ideally, a vulnerability assessment will capture dynamic trends and processes (per Figure 2), not just a snapshot. And the relationship is circular: high levels of vulnerability mean that a population is particularly at risk to the negative impact of drought. In turn, the impact of a prolonged drought may erode the asset base of that population, leaving them more vulnerable to future drought events in the absence of mitigating or preparedness measures.

Although we can do little if anything to alter drought occurrence, there *are* things we can do to reduce vulnerability. This is where government policy comes into play. For example, underlying vulnerability can be reduced through development programs targeted to the poorest, to strengthen their asset base. Government can provide relief from the immediate impact of drought through livestock support and provision of subsidized or free food. In the drought-prone districts of northern Kenya, for example, relief programs have ranged from emergency livestock purchase schemes (designed to protect pastoralists' purchasing power in the face of drought and reduce pressure on grazing resources) to free food distribution, to food or cash-for-work programs. The effectiveness of these relief and recovery programs depends on issues such as the timeliness of the intervention, the scale and adequacy of resources, and the approach to targeting. Implemented well, they have played a key role in protecting both lives and livelihoods (see, for example, Buchanan-Smith and Barton, 1999; Buchanan-Smith and Davies, 1995).

IV. THE CHALLENGE OF DROUGHT EARLY WARNING

Although an understanding of underlying vulnerability is essential to understand the risk of drought in a particular location and for a particular group of people, a drought early warning system (DEWS) is designed to identify negative trends and thus to predict both the occurrence and the impact of a particular drought *and* to elicit an appropriate response (Buchanan-Smith and Davies, 1995).

Numerous natural indicators of drought should be monitored routinely to determine drought onset, end, and spatial characteristics. Severity must also be evaluated continuously on frequent time steps. Although droughts originate from a deficiency of precipitation, it is insufficient to rely only on this climate element to assess severity and resultant impacts. An effective DEWS must integrate precipitation data with other data such as streamflow, snowpack, groundwater levels, reservoir and lake levels, and soil moisture in order to assess drought and water supply conditions (see Chapter 3).

These physical indicators and climate indices must then be combined with socioeconomic indicators in order to predict human impact. Socioeconomic indicators include market data—for example, grain prices and the changing terms of trade between staple grains and livestock as an indicator of purchasing power in many rural communities—and other measures of coping strategies. Poor people usually employ a sequence of strategies in response to drought. Early coping strategies rarely cause any lasting damage and are reversible. In many poor rural communities, examples of early coping strategies include the migration of household members to look for work, searching for wild foods, and selling nonproductive assets. If the impact of the drought intensifies, these early strategies become unviable and people are forced to adopt more damaging coping strategies, such as selling large numbers of livestock or choosing to go hungry in order to preserve some productive assets. Once all options are exhausted, people are faced with destitution and resort to crisis strategies such as mass migration or displacement (Corbett, 1988; Young et al., 2001). Monitoring these coping strategies provides a good indicator of the impact of drought on the local population, although by the time there is evidence of the later stages of coping, it is usually too late to launch a preventative response.

Effective DEWSs are an integral part of efforts worldwide to improve drought preparedness. (Many DEWSs are, in fact, a subset of an early warning system with a broader remit—to warn of other natural disasters and sometimes also conflict and political instability.) Timely and reliable data and information must be the cornerstone of effective drought policies and plans. Monitoring drought presents some unique challenges because of the hazard's distinctive characteristics.

An expert group meeting on early warning systems for drought preparedness, sponsored by the World Meteorological Organization (WMO) and others, recently examined the status, shortcomings, and needs of DEWSs and made recommendations on how these systems can help in achieving a greater level of drought preparedness (Wilhite et al., 2000b). This meeting was organized as part of WMO's contribution to the Conference of the Parties of the U.N. Convention to Combat Desertification (UNCCD). The proceedings of this meeting documented recent efforts in DEWSs in countries such as Brazil, China, Hungary, India, Nigeria, South Africa, and the United States, but also noted the activities of regional drought monitoring centers in eastern and southern Africa and efforts in West Asia and North Africa. Shortcomings of current DEWSs were noted in the following areas:

- *Data networks*—Inadequate station density, poor data quality of meteorological and hydrological networks, and lack of networks on all major climate and water supply indicators reduce the ability to represent the spatial pattern of these indicators accurately.
- *Data sharing*—Inadequate data sharing between government agencies and the high cost of data limit the application of data in drought preparedness, mitigation, and response.
- *Early warning system products*—Data and information products are often too technical and detailed. They are not accessible to busy decision makers who, in turn, may not be trained in the application of this information to decision making.
- *Drought forecasts*—Unreliable seasonal forecasts and the lack of specificity of information provided by forecasts limit the use of this information by farmers and others.
- *Drought monitoring tools*—Inadequate indices exist for detecting the early onset and end of drought, although the Standardized Precipitation Index (SPI) was cited as an important new monitoring tool to detect the early emergence of drought.
- *Integrated drought/climate monitoring*—Drought monitoring systems should be integrated and based on multiple physical *and* socioeconomic indicators to fully understand drought magnitude, spatial extent, and impacts.
- *Impact assessment methodology*—Lack of impact assessment methodology hinders impact estimates and the activation of mitigation and response programs.

- *Delivery systems*—Data and information on emerging drought conditions, seasonal forecasts, and other products are often not delivered to users in a timely manner.
- *Global early warning system*—No historical drought database exists and there is no global drought assessment product that is based on one or two key indicators, which could be helpful to international organizations, nongovernmental organizations (NGOs), and others.

As has now been well documented (see, for example, Buchanan-Smith and Davies, 1995), early warning alone is not enough to improve drought preparedness. The key is whether decision makers listen to the warnings and act on them in time to protect livelihoods before lives are threatened. There are many reasons why this is often the "missing link." For example, risk-averse bureaucrats may be reluctant to respond to predictions, instead waiting for certainty and quantitative evidence. This invariably leads to a late response to hard evidence that the crisis already exists. Who "owns" the early warning information is also critical to how it is used. Does it come from a trusted source, or is it treated with suspicion? Ultimately, sufficient political will must exist to launch a timely response and hence to heed the early warnings (Buchanan-Smith and Davies, 1995).

V. EXAMPLES OF THE INTERACTION OF DROUGHT WITH THE WIDER SOCIAL/POLITICAL CONTEXT

A. Southern Africa Food Crisis of 2002–2003

Initially, the southern Africa crisis was presented as a food crisis triggered by drought. But it soon became apparent that the roots of the crisis were much more complex. The declines in rainfall and crop production were not as great as the drought of a decade earlier, yet the food crisis in 2002–2003 seemed to be more serious. The reasons varied from one country to another. Mismanagement and poor governance were key factors in Zimbabwe, Zambia, and Malawi. In Malawi, one manifestation of this was the well-documented mismanagement of strategic grain reserves in the early stages of the food crisis. In Zimbabwe, economic decline had been dramatic, with a 24% decline in gross domestic product (GDP) over 3 years to 2003, rising unemployment, hyperinflation, and shortages of foreign currency with consequent shortages of basic commodities such as cooking oil, maize meal, and fuel (Cosgrave et al., 2004). In both Zimbabwe and Angola, conflict and/or political instability have been significant contributing factors. Generally, there has been a failure of growth in this region in recent years and rising poverty. This has been associated with market failure in the context of market liberalization (Ellis, 2003). In addition, the high prevalence of HIV and AIDS has had a devastating impact at the household level across the region in ways that are only beginning to be understood (see, e.g., De Waal, 2002, on the controversial "new variant famine" theory). According to estimates, 20–25% of working-age adults in the region are now infected with HIV or AIDS.

Thus, the drought of the early 2000s visited a population that was poorer and more vulnerable than it had been 10 years earlier. Although the drought was less severe in meteorological terms, the combined effect of drought plus these other factors meant that the impact was much more devastating. It is probably fair to say that the extent to which vulnerability had increased was not fully appreciated until the food crisis deepened and in-depth studies and analyses were carried out (Cosgrave et al., 2004). Ellis's (2003) study of human vulnerability in southern Africa concludes:

> The main policy implication ... is that the contemporary donor-government framework (PRSPS and decentralization) needs substantial re-thinking and strengthening concerning the facilitating institutional context that it manifestly fails to specify, that would give people space to thrive rather than just survive. (p. iii)

B. Drought and War in South Sudan in 1998

Bahr El Ghazal province in South Sudan suffered from a devastating famine in 1998 (Buchanan-Smith, 2003b). The total number of people who died is unknown. What is known is that malnutrition and mortality rates were among the highest ever recorded. The causes of the famine were a combination of drought and conflict. The civil war in South Sudan dates back some 40 years, and this particular province has been subjected to more than a decade of violence and looting. The 3-year drought leading up to 1998, caused by El Niño, further depleted the asset base of an already impoverished population. Coping strategies were severely stretched. However, it was the combination of drought and a number of war-related incidents that created the famine. In 1997, there had been a fundamental shift in the war economy of the region as the Sudan People's Liberation Army (SPLA) made military advances, driving out government troops and causing a collapse in trading relationships between garrisons and the local population, thus depriving the latter of a key source of livelihood. One of the final blows for the urban population in Bahr El Ghazal were the attacks on three garrison towns, causing massive displacement of approximately 130,000 people from towns into rural areas that were already facing acute food insecurity. Just when the food security situation became cataclysmic, the Sudan Government imposed a ban on all relief flights in early February 1998, supposedly for security reasons. When the ban was partially lifted in early March, relief flights were allowed into only four sites, creating a fatal magnet effect on an already highly stressed population. The concentration of people resulted in increased transmission of disease, and rocketing mortality.

The irony of this case is that early warning of a crisis was provided, but it was not acted on until it was too late. The challenge for early warning practitioners was to convince decision makers that "this year was different" as a result of the complex interaction of factors that ultimately created such deadly conditions. Although a number of aid workers warned that 1998 could be the worst year for a decade, by their own admission they did not foresee the speed at which famine would develop (Buchanan-Smith, 2003b). Just as no two droughts are the same, no two famines are the same. Thus, early warning is an art, not a science.

C. Recent Drought Years in the United States, 1996–2004

Both the micro and macro contexts are also important in developed countries such as the United States, Canada, and Australia, where recent droughts have resulted in widespread and severe impacts in many sectors. In these instances, greater institutional capacity and resources are available to monitor, prepare for, and respond to drought, but the impacts are still devastating to livelihoods as well as to the environment and social fabric. In the United States, recent droughts have produced far-reaching impacts on many economic sectors while also resulting in serious social hardships, especially in the agricultural community, and significant environmental consequences. Impact estimates for the 2002 drought exceeded $20 billion, although there has not been a systematic assessment conducted at the national level. Average annual losses from drought in the United States have been estimated at $6–8 billion (FEMA, 1995). Government actions often lead to drought relief appropriations in the billions of dollars. Recent droughts have stimulated greater interest in drought preparedness and mitigation by states and also by the federal government as a method to reduce vulnerability and impacts. These drought years have had substantial impacts on agriculture, forest fires, transportation, recreation and tourism, and energy production. In the west and parts of the southeast, population growth has been dramatic in recent years and has furthered the debate on the sustainability of these growth rates in light of the occurrence of consecutive drought years. Water resources are often overappropriated in many parts of the western United States, so the provision of water in support of agriculture and other uses is becoming more contentious for much of this region.

VI. DROUGHT-VULNERABLE VS. DROUGHT-RESILIENT SOCIETY

The Drought Discussion Group of the International Strategy for Disaster Reduction (ISDR) has proposed a new paradigm to improve understanding of the drought hazard in the macro and micro contexts with the goal of enhancing drought preparedness and mitigation efforts in all settings ranging from local to national and from developing to developed countries (ISDR Drought Discussion Group, 2003). This new paradigm emphasizes greater understanding and description of both the physical features of the hazard and the social factors that influence societal vulnerability. Figures 3 and 4 are modified from the Drought Discussion Group's report and represent the characteristics of a drought-vulnerable society (i.e., crisis management) and the discussion group's vision for future drought management efforts, respectively.

The society portrayed in Figure 3 is vulnerable to drought and has not developed the institutional capacity to monitor its onset and end, to mitigate risk, or to launch a timely relief response. In this example, there has been no vulnerability assessment of who and what is at risk and why, a fundamental prerequisite of a risk-based approach to drought management. The result is a reactive approach to drought management characterized by delayed crisis response in the post-drought setting. This often leads to far-reaching negative impacts and a long period of recovery. Often another drought episode will occur before the recovery process is complete.

Under the new paradigm for a drought-resilient society, a risk-based drought policy is developed with preparedness plans and proactive mitigation strategies. It is part of a long-term management strategy directed at reducing societal vulnerability to drought. Security is a prerequisite. A comprehensive early warning system that integrates a wide range of physical and social indicators has been developed and implemented. The early warning system works well in delivering time-sensitive information to decision makers, who, in turn, have the political will and resources to apply this information as part of a comprehensive risk-reducing strategy. In this model, governance systems work and vulnerable people are able to claim their rights. Although this may be an ideal, it highlights the conditions necessary to reduce the risk of drought and the necessary role of government in drought mitigation and management.

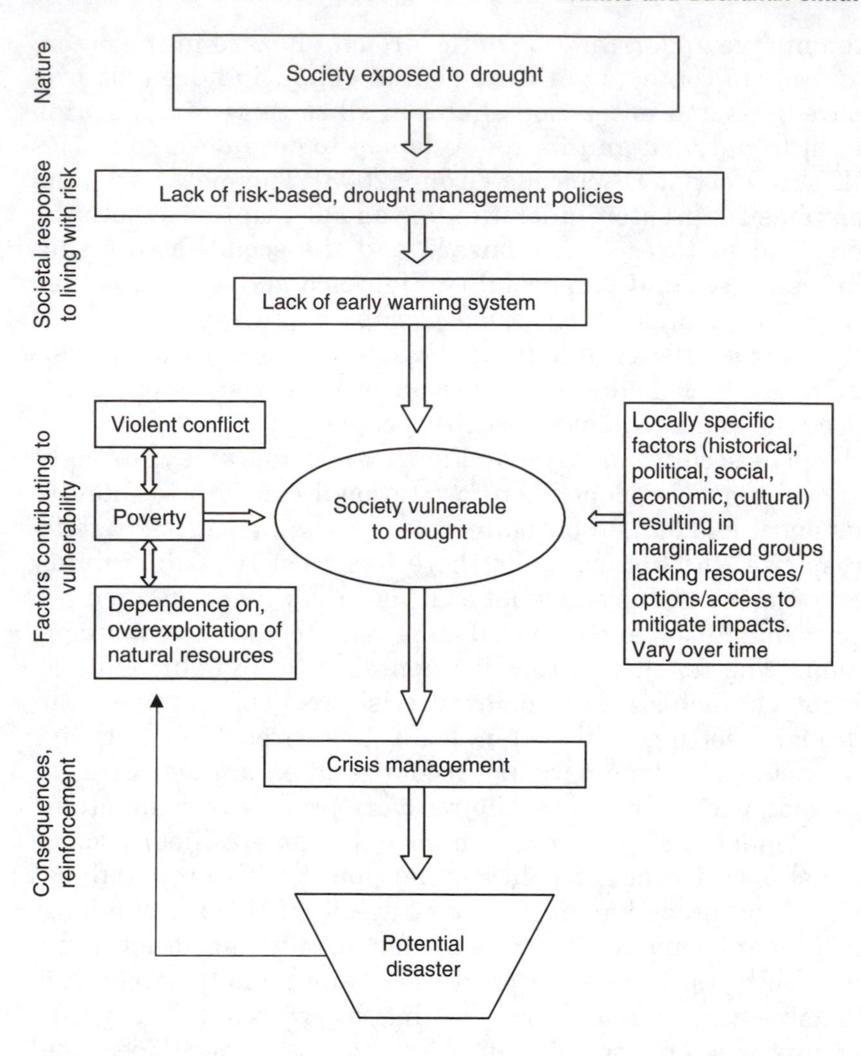

Figure 3 Drought-vulnerable society. (Modified from ISDR Discussion Group.)

VII. SUMMARY AND CONCLUSION

Drought is an insidious natural hazard that is a normal part of the climate of virtually all regions. It should not be viewed as merely a physical phenomenon. Rather, drought is the result of the interplay between a natural event and the

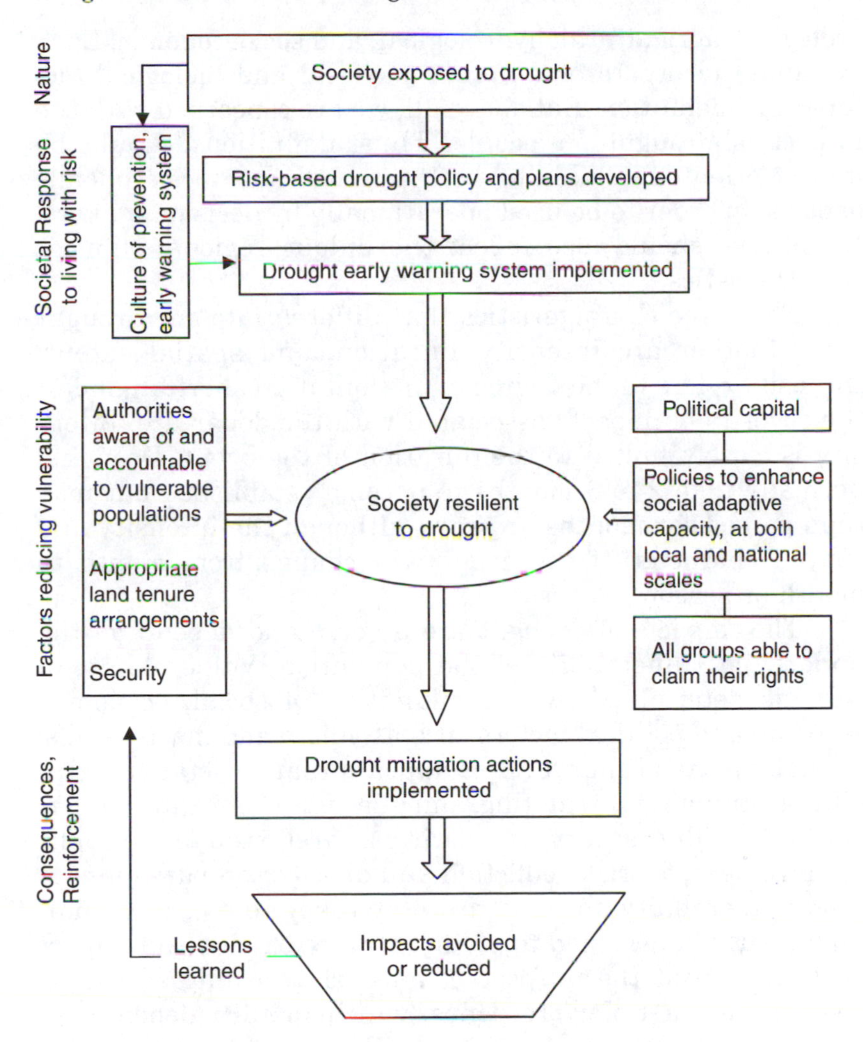

Figure 4 Drought-resilient society. (Modified from ISDR Discussion Group.)

demand placed on a water supply by human-use systems. It becomes a disaster if it has a serious negative impact on people in the absence of adequate mitigating measures.

Many definitions of drought exist; it is unrealistic to expect a universal definition to be derived. Drought can be grouped by type of disciplinary perspective as follows: meteorological, agricultural, hydrological, and socioeconomic. Each discipline incorporates different physical and biological factors in its definition. But above all, we are concerned with the impact of drought on people. Thus, definitions should be impact oriented, combining both physical and socioeconomic aspects, in order to be used operationally by decision makers. Definitions should also reflect the unique regional climatic characteristics.

The three characteristics that differentiate one drought from another are intensity, duration, and spatial extent. Intensity refers to the degree of precipitation shortfall and/or the severity of impacts associated with the departure. Intensity is closely linked to the duration of the event. Droughts normally take 2 to 3 months to become established but may then persist for months or years, although the intensity and spatial character of the event will change from month to month or season to season.

The impacts of drought are diverse and depend on the underlying vulnerability of the population. Vulnerability, in turn, is determined by a combination of social, economic, cultural, and political factors, at both micro and macro levels. In many parts of the world, it appears that societal vulnerability to drought is escalating, and at a significant rate. Understanding vulnerability is a critical first step in drought management, for risk reduction and disaster preparedness. A good vulnerability analysis can play a key role in underpinning a DEWS, designed to predict the occurrence and impact of drought. But the purpose of an early warning system is also to elicit a timely response, which in turn depends on adequate resources and political will.

It is imperative that increased emphasis be placed on mitigation, preparedness, and prediction and early warning if society is to reduce the economic and environmental damages associated with drought and its personal hardships. This will require interdisciplinary cooperation and a collaborative effort with policy makers at all levels.

REFERENCES

Blaikie, P, T Cannon, I Davis, B Wisner. *At Risk: Natural Hazards, People's Vulnerability, and Disasters.* London: Routledge Publishers, 1994.

Buchanan-Smith, M. Case Study of Gujarat, India—The Earthquake and Communal Rioting: 2001–2002. Prepared for the BRCS-hosted seminar on Natural Disasters and Complex Political Emergencies, 6th November 2003, London, 2003a.

Buchanan-Smith, M. Case Study of the Bahr El Ghazal Famine, South Sudan: 1997–1998. Prepared for the BRCS-hosted seminar on Natural Disasters and Complex Political Emergencies, 6th November 2003, London, 2003b.

Buchanan-Smith, M, D Barton. Evaluation of the Wajir Relief Programme: 1996–1998. Oxford, UK: Oxfam, 1999.

Buchanan-Smith, M, S Davies. *Famine Early Warning and Response—The Missing Link.* London: IT Publications, 1995.

Chambers, R. *Vulnerability: How the Poor Cope.* IDS Bulletin 20(2). Sussex, UK: Institute of Development Studies, 1989.

Collinson, S, ed. *Power, Livelihoods and Conflict: Case Studies in Political Economy Analysis for Humanitarian Action.* HPG Report 13. London: Overseas Development Institute, 2003. Available on line at www.odi.org.uk/hpg/papers/hypgreport13_a.pdf and www. odi.org.uk/hpg/papers/hpgreport13_b.pdf.

Corbett, J. Famine and household coping strategies. *World Development* 16(9):1092–1112, 1988.

Cosgrave, J, A Jacobs, M McEwan, P Ntata, M Buchanan-Smith. A Stitch in Time? Independent Evaluation of the Disasters Emergency Committee's Southern Africa Crisis Appeal. July 2002 to June 2003. Volume 1: Main Report. Oxford, UK: Valid International, 2004.

De Waal, A. New Variant Famine in Southern Africa. Presentation for SADC VAC Meeting, Victoria Falls, Zimbabwe. 17–18 October, 2002.

Ellis, F. Human Vulnerability and Food Insecurity: Policy Implications. Forum for Food Security in Southern Africa, August, 2003.

FEMA. *National Mitigation Strategy.* Washington, D.C.: Federal Emergency Management Agency, 1995.

Gillette, HP. A creeping drought under way. *Water and Sewage Works* March:104–105, 1950.

Glantz, MH. *Climate Affairs: A Primer.* Washington, D.C.: Island Press, 2003.

IPCC. *Climate Change. The Scientific Basis. Contributions of Working Group I to the Third Assessment Report of the Intergovernmental Panel on Climate Change.* JT Houghton, Y Ding, DJ Griggs, M Noguer, PJ van de Linden, X Dai, K Maskell, CA Johnson, eds. U.K. and New York: Cambridge University Press, 2001.

ISDR Drought Discussion Group. Drought: Living with Risk—An Integrated Approach to Reducing Societal Vulnerability to Drought. Secretariat, International Strategy for Disaster Reduction, Geneva, Switzerland, 2003.

O'Neill, WG. An Introduction to the Concept of Rights-Based Approach to Development: A Paper for InterAction. Washington, D.C.: Interaction, December, 2003. Available online at www.inte-action.org/files.cgi/2495_RBA_1-5-04.pdf.

Pessoa, D. Drought in Northeast Brazil: Impact and Government Response. In: DA Wilhite, WE Easterling, eds. *Planning for Drought: Toward a Reduction of Societal Vulnerability* (pp. 471–488). Boulder, CO: Westview Press, 1987.

Save the Children (U.K.). The Household Economy Approach: A Resource Manual for Practitioners. London: Save the Children, 2000.

Tannehill, IR. *Drought: Its Causes and Effects.* Princeton, NJ: Princeton University, 1947.

Wilhite, DA, ed. *Drought: A Global Assessment*. London: Routledge Publishers, 2000.

Wilhite, DA, MH Glantz. Understanding the drought phenomenon: The role of definitions. *Water International* 10:111–120, 1985.

Wilhite, DA, MJ Hayes, C Knutson, KH Smith. Planning for drought: Moving from crisis to risk management. *Journal of the American Water Resources Association* 36:697–710, 2000a.

Wilhite, DA, MKV Sivakumar, DA Wood, eds. Early Warning Systems for Drought Preparedness and Management. Proceedings of an Experts Meeting, World Meteorological Organization, Geneva, Switzerland, 2000b.

Young, H, S Jaspars, R Brown, J Frize, H Khogali. *Food-Security Assessments in Emergencies: A Livelihoods Approach.* HPN Network Paper No. 36. London: ODI, 2001.

Part II

Drought and Water Management: The Role of Science and Technology

2

The Challenge of Climate Prediction in Mitigating Drought Impacts

NEVILLE NICHOLLS, MICHAEL J. COUGHLAN, AND KARL MONNIK

CONTENTS

I. FORECASTING DROUGHT

A. Introduction

Examination of the long-term climate records in some regions around the globe reveals persistent trends and periods of below-average rainfall extending over years to a decade or more, while other regions exhibit episodic, shorter droughts. Hence it is useful to consider the prediction of droughts on seasonal to interannual timescales and, separately, on longer decadal timescales.

B. Seasonal to Interannual Prediction

Our theoretical ability to make an explicit, reliable prediction of an individual weather event reduces to very low levels by about 10–15 days (this is called the "weather predictability barrier"), so forecasts with lead times longer than this should be couched in probabilistic terms. Consequently, a forecast with a lead time of a month or more requires a statistical basis for arriving at a set of probability estimates for the ensuing seasonal to interannual conditions. Two approaches allow us to derive these estimates. The first is based on statistical analyses of the climatic record and assumptions about the degree to which the statistics of the future record will differ from the past record. The second, and more recent, approach is based on the generation of statistics from multiple, explicit predictions of weather conditions using computer models of the climate system.

1. Forecasts Based on Empirical Analysis of the Climate Record

The fact that the earth's climate system is driven primarily by the regular rotation of the earth around the sun led to many efforts during the last two centuries to link the recurrence of droughts with cycles observed in the movements and features of heavenly bodies. Notable among these efforts were schemes based on the phases of the moon and the occurrence of sunspots. These purported linkages have been proven to be statistically insignificant, evanescent, or of little practical value. Nonetheless, there *are* recurring climate patterns, caused by the interacting dynamics of the earth's atmosphere and oceans, that provide some scope for prediction. The development of comprehensive climate records and the growth of computing power over the past 20 years or so have enabled a wide range of powerful statistical tools to be brought to bear to tease out these patterns and incorporate them into empirical algorithms for predicting future seasonal patterns.

One of the earliest identified and most powerful of these rhythms, apart from the annual cycle itself, is the El Niño/Southern Oscillation phenomenon, often referred to as ENSO. The robustness of ENSO-related patterns over time in the distribution of rainfall, air and sea temperatures, and other climatic variables, and the fact that the phenomenon is caused by slowly varying components of the ocean–atmosphere system, renders it useful as a predictor. ENSO-based indices (e.g., Troup, 1965; Wolter and Timlin, 1993) are the dominant predictors for statistically based seasonal prediction schemes over many parts of the globe, although other indices are now being combined with ENSO for different regions—for example, North Australia/Indonesia (Nicholls, 1984), the Indian Ocean (Drosdowsky, 1993), and the North Atlantic (McHugh and Rogers, 2001).

One of the simplest of the statistical prediction methods is based on the underlying premise that the behavior of a dominant pattern in the future climate will continue to replicate the behavior observed in the past record. A systematic scan of the record of the Southern Oscillation Index (SOI), for example, can reveal occurrences, or "analogs," when the track of the index over recent months was "similar" to the track in corresponding months in several past years (Stone and Auliciems, 1992).

More complex approaches for deriving empirically based forecasting schemes have been implemented in several operational forecasting centers throughout the world. A typical example is the methodology developed for the scheme used by the Australian National Climate Centre for forecasting probability ranges of seasonal (3-month) rainfall and temperatures (maximum and minimum). This methodology (Drosdowsky and Chambers, 1998) involves:

1. Identification of predictands (e.g., rainfall and temperature) and possible predictors (sea surface temperatures representative of one or more areas).
2. Construction of the statistical model, including procedures for the optimum selection and weighting of predictors.
3. Verification or estimation of forecast skill.

Improvements in the forecast skill of such statistical schemes likely will plateau, because they are generally constrained by a limited number of useful predictors and relatively short periods of data. Most statistical methods also exhibit large variations in their skill level throughout the year—because of seasonal variations in statistical relationships between climate variables—and for particular regions. Further, if there are slow or even rapid changes of climate underway that are not adequately captured in the past record (as has indeed occurred in recent decades), it is possible that the skill of the forecasts may be lower than would be the case in a more stable climate. Despite these problems, statistically based schemes will likely remain useful and sometimes potent weapons for forecasting meteorological droughts.

2. Explicit Computer Model Predictions

Between about 1970 and 1980, the basis for generating daily weather forecasts moved from sets of empirical, observationally based rules and procedures to explicit predictions made by computer models of the three-dimensional structure of the atmosphere. However, in order to make similar progress in computer-based forecasting on longer time scales, it was essential to incorporate the slower contributions to variability from ocean circulations and variations of the land surface. In the last two decades, there have been significant improvements in the understanding of processes in the atmosphere and the ocean and in the way in which the atmosphere interacts with, or is coupled to, the various underlying surfaces. These advances in knowledge, combined with an expanded range of data and a massive increase in computer power, have made it possible to develop prediction schemes based on computer models that represent the entire earth/ocean/atmosphere system (e.g., Stockdale et al., 1998).

Although such schemes are still in their infancy, rapid developments are underway. For example, it is now evident that the details of a season's outcome are modulated by processes occurring on shorter, intraseasonal timescales, which may affect, for example, the timing and intensity of patterns of decreased or increased rainfall (Slingo et al., 1999; Schiller and Godfrey, 2003). Hence, efforts are being made to ensure that computer models of the coupled system can simulate and predict such short-term modes of variability. It is likely, too, that improvements in predictive skill on seasonal to interannual timescales, and hence improvements in prediction of droughts, will be realized from further expansions in the observational base, especially from the oceans (e.g., Smith, 2000); from the ability to generate larger prediction ensembles from individual computer models (Kumar and Hoerling, 2000); and from combined ensembles from several different computer models (Palmer et al., 2004).

Work is also underway to improve the spatial resolution at which seasonal forecasts can be made, through statistical "downscaling" techniques, through the nesting of high-resolution regional-scale climate models within coarser resolution global-scale models, and by increasing the resolution of the global models.

Despite these developments, it will never be possible to consistently generate forecasts of individual events beyond the 10–15-day weather predictability barrier. What these developments promise, however, is the generation of reliable short-term model-based "forecast climatologies" from which one can then generate probabilistic assessments of likely climate anomalies over a month, a season, or longer—for example, of conditions conducive to the onset, continuation, or retreat of drought.

C. Can We Forecast Droughts on Even Longer Time Scales?

Improvements in seasonal forecasting have arisen from advances in knowledge made as a result of the careful analysis of data collected over time. The growth in knowledge about the circulation of the oceans and its modes of variability, which was stimulated in large measure during the 1980s with the implementation of the Tropical Ocean Global Atmosphere (TOGA) and World Ocean Circulation Experiment (WOCE) projects of the World Climate Research Program, is beginning to reap rewards in the identification and understanding of even slower modes of variability than are at work on seasonal timescales. In particular, in the two ocean basins that extend to both polar regions, evidence exists in both oceanic and atmospheric records of quasi-rhythmic variations with timescales of a decade or so known as the North Atlantic Oscillation (Hurrell, 1995) and the Pacific Decadal Oscillation (Nigam et al., 1999). There is also evidence of decadal variations in ENSO. Its signal, for example, has been more evident in rainfall patterns of the western regions of the United States since the late 1970s compared to the previous quarter century, when its influence was stronger over southern and central regions (Rajagopalan et al., 2000). Slow variations of this nature complicate the challenge of forecasting drought using the statistics of the historical record alone.

Much has yet to be learned about what drives these slow variations (Miller and Schneider, 2000; Alexander et al., 2001) and thence how to predict them. We must continue to advance our knowledge in this area if we are to improve our skill in forecasting drought, especially in those areas that have seen downward trends in rainfall—for example, the Sahel region of West Africa (Zeng et al., 1999) and the far southwest of Western Australia (IOCI, 2002).

The path to better prediction of droughts on the decadal scale involves identifying correlated patterns of variability in atmospheric and oceanic records, investigating the physical and dynamic processes at work, representing those processes within a hierarchy of computer models, and developing sets of statistics from a range of predictive models. Although research tends to focus on one scale or the other, implementation of the results at the practical level must integrate the outcomes of many complex processes across all timescales. This will be best done by those models of the coupled system that have the capacity to represent all the key processes involved, whatever the timescale. This is clearly not a trivial task.

II. CLIMATE PREDICTION AND DROUGHT EARLY WARNING SYSTEMS

Early warning systems (EWSs) have become increasingly successful at recognizing the development of potential famines and droughts. Saidy (1997) pointed out that in 1992 EWSs were successful in sounding the alarms about the drought emergency. Although some warnings, such as those given in southern Africa during 1997–1998, were not followed by full-blown droughts and famines, such events are not necessarily forecast failures because most, if not all, seasonal forecasts are issued as probabilities for dry, near-normal, or wet conditions. Although there has been increasing focus on economic and social indices to complement physical information, a seasonal forecast for drought potentially provides an early indication of impending conditions. Economic and social indices tend to follow the development of drought and are valuable to confirm the existence of drought conditions.

Food security will exist when all people, at all times, have access to sufficient, safe, and nutritious food for a healthy and active life (World Food Summit, 1996). However, certain parts of the globe have shown themselves to be more vulnerable to droughts and famines because of variable climate, marginal agriculture, high dependence on agriculture, and social and military conflict. The populations of many countries in sub-Saharan Africa suffer from chronic malnutrition, with frequent famine episodes. Achieving food and water security will remain a development priority for Africa for years to come. Even in a nation that is food secure at the national level, household food security is not guaranteed.

A "famine EWS" has been defined as a system of data collection to monitor people's access to food (Buchanan-Smith, 1997). However, this definition suggests the collection of monitoring data is sufficient to provide an early warning. The provision of prediction information (a forecast) increases the time available to elicit a response, but it does not guarantee that the appropriate response will result. A famine EWS should consider the demand side (what is required), the supply side (what is available), and food entitlement (the ability to access what is available). Drought early warning plays an important role in forecasting the supply side.

Before too much investment of time and effort is placed in drought or rainfall early warning (as a physical event), one needs to ask what the "drought early warning system" is intended to achieve. A drought early warning forecast must identify components of a drought that strongly affect food supply and the development of famine conditions, along with factors affecting water supply. Drought EWSs should incorporate a broad range of information in order to provide a balanced perspective of conditions. Although no particular kind of information is a unique indicator, a famine EWS cannot do without physical information such as rainfall (including forecasts) or drought early warning. In fact, these types of information are practically the only types that can provide a longer lead-time forecast to the development of a drought.

Glantz (1997) defined famine as "a process during which a sharp decline in nutritional status of at-risk population leads to sharp increases in mortality and morbidity, as well as to an increase in the total number of people at risk." Quoting Murton (1991), he goes on to say that the purpose of an early warning system is "to inform as many people as possible in an area-at-risk that a dangerous and/or damaging event is imminent and to alert them to actions that can be taken to avoid losses."

The first purpose of a drought EWS is to determine the probability of a drought event and to monitor its spatial extent, duration, severity, and those who may be potentially affected. This requires an appreciation of the climatology of the area and the crop calendars. As described by Walker (1989), a famine EWS should detect, evaluate, and predict the hazard. It uses monitoring tools such as remote sensing, market conditions, and climate forecasts, as well as geographical information systems to isolate the extent of the hazard area. Huss-Ashmore (1997) examined the question of what predictions are needed for a famine EWS. In order to pursue an increase in food imports at a national level, governments require a significantly earlier indicator of potential problems. However, information such as drought early warning indicates only the *potential* for problems, whereas output-related indicators show the emergence of *actual* problems. Delaying a response until this information is available would generally result in some level of food shortage.

A significant challenge in developing a drought EWS is the range of spatial and temporal scales of the information available. On one hand, market prices of staple crops on a week-to-week basis may be monitored. But this information needs to be integrated with global three-monthly (and even possibly longer) regional climate forecasts. Related to this problem is information that only partly reflects the real information requirement. For example, global climate forecasts generally forecast seasonal rainfall totals, but this information may not relate to the necessary agricultural rainfall distribution during the season or the required crop growth season.

It is important to ensure that the information is used to the best advantage in order to determine a timely and appropriate response. Walker (1989) noted that this involves interpreting the available information and preparing a message that is clear and easily understood. To realize the benefits of early warning, response is the issue, not developing ever-more sophisticated indicators (IFRC, 1995). This requires careful interpretation and presentation of the data. Bulletins such as those prepared by the FEWS NET, Southern African Development Community Food, Agriculture and Natural Resources Vulnerability Assessment Committee, and World Food Program make use of maps, tables, diagrams, and short paragraphs of text to get the message across. Products are tailored to target groups such as government ministers, donors, humanitarian organizations, and disaster management authorities. Walker (1989) highlighted the need to spread the message through the appropriate channels in order to elicit the appropriate response. Wilhite (1990) emphasized the need for an EWS to provide decision makers at all levels with information concerning the onset, continuation, and termination of drought conditions—essential for formulating an adequate response to an expected drought situation. Saidy (1997) suggested that the early warning units be connected to response mechanisms and functionally be responsible for early warning and response. This would benefit both those who prepare the early warning bulletins and those in charge of response.

Different types of information are ready at different times. Climate forecasts may provide indications of a drought several months in advance, whereas social and economic indicators will gain prominence at the stage when the drought or famine sets in. Sometimes anecdotal information and media reports can provide early warnings. Good baseline data is essential because many areas regularly experience pre-harvest "hungry seasons," so an indicator that simply highlights a seasonal event is not useful. A drought EWS needs to include all components that could contribute to a drought or a drought-related famine. This includes production (weather, yield, carry-over stocks), exchange (markets, prices, and availability), consumption (affordability, health) of food, and communication. A broader range of indicators can result in a more robust index of drought or famine. Many EWSs now use multi-indicator models that incorporate a wide range of biophysical and socioeconomic indicators (Buchanan-Smith, 1997).

A vulnerability analysis should complement a drought EWS. This could indicate areas that will be first affected and help with prioritization of humanitarian aid. Matching the impending hazard with the vulnerability of farming systems and rural communities enables decision makers to tailor response strategies for the greatest impact. A vulnerability profile should include, inter alia, trends in recent rainfall, production, prices, reserves, nutritional status, soil fertility, and household status (Ayalew, 1997)

For many years, the primary purpose of drought EWSs has been, directly or indirectly, to notify external organizations of the impending adverse situation. The traditional focus for assistance for African countries has been western countries and international aid organizations. Often external donor aid is driven by scenes of devastation. Thus the very act of responding in good time to a drought warning or potential drought situation may lead to a decrease in response. To encourage the long-term sustainability of drought EWS organizations, they need to integrate the outlooks with farming strategies the local population can use to decrease their inherent vulnerability. Examples of such practices include the increase of rainfall harvesting technology and the use of an "outlook spreadsheet." Developed by E. Mellaart (personal communication, 2002), the outlook spreadsheet allows farmers to examine potential yield or economic profit under various climate and farming system regimes. The user enters into the model the current seasonal forecast and then determines what the yield (or economic profit/loss) might be, depending on the agricultural choices made and the range of possible weather outcomes, either for a single season or over several seasons. Yields can be estimated assuming that the forecast is correct or is completely wrong, or when a risk-reducing strategy is adopted. The spreadsheet needs to be seeded with yield (or economic) data for a range of management options and under a range of weather scenarios. This could provide a useful focus for agricultural research.

Monitoring and analysis of weather systems must remain a central part of EWSs. Early warning systems have played a critical role in identifying and alerting key decision makers to imminent droughts. However, as they mature, the emphasis will no doubt have to switch to a greater extent to domestic applications.

III. IMPEDIMENTS TO USING CLIMATE PREDICTIONS FOR DROUGHT MITIGATION

A survey of the scientific literature, and experience in operational seasonal climate prediction, reveals that a variety of impediments obstructs the optimal use of seasonal climate forecasts, especially in drought mitigation (Nicholls, 2000).

The limited skill obtainable with climate predictions is well known and is often cited as a reason for the limited use of climate predictions. Awareness of the existence of an El Niño episode in 1997 led to mitigation efforts in southern Africa in anticipation of a possible drought in 1998. A major drought did not materialize that year; so the forecast led to preparations that created negative impacts, such as reducing the amount of seed purchased by farmers because they feared their crops would fail (Dilley, 2000).

Glantz (1977) noted a variety of social, economic, environmental, political, and infrastructural constraints that would limit the value of even a perfect drought forecast. He concluded that a drought forecast might not be useful until adjustments to existing social, political, and economic practices had been made. Hulme et al. (1992), in a study of the potential use of climate forecasts in Africa, suggested that forecasts may be useful at the national and international level (e.g., in alerting food agencies to possible supply problems), but they also concluded that improvements in institutional efficiency and interaction are needed before the potential benefits of the forecasts could be realized. Broad and Agrawala (2000), discussing the 2000 food crisis in Ethiopia, concluded that “even good climate forecasts are not a panacea” to the country’s food crisis.

Felts and Smith (1997) noted that many decision makers receive climate information through secondary sources, such as the popular media or professional or trade journals, rather than from primary sources such as meteorological agencies. Nicholls and Kestin (1998) discussed the communication problems associated with the Australian Bureau of Meteorology’s seasonal climate outlooks during the 1997–1998 El Niño.

Toward the end of 1997 it became clear that there was a wide gap between what the bureau was attempting to say (i.e., an increased likelihood of drier-than-normal conditions) and the message received by users (i.e., definitely dry conditions, perhaps the worst drought in living memory). Some of this gap arose from confusion about the use of terms such as *likely* in the outlooks. It appears that users and forecasters interpret *likely* in different ways (Fischhoff, 1994). Those involved in preparing the forecasts and media releases intended to indicate that dry conditions were more probable than wet conditions. Many users, however, interpreted *likely* as "almost certainly dry, and even if it wasn't dry then it would certainly not be wet."

Users may tend to underreact to a forecast or downplay the likelihood of disasters (Felts and Smith, 1997). At a policy level, one might assume that potential users of climate forecasts might be more knowledgeable about the basis and accuracy of climate prediction, and its potential value, compared with the average individual user such as a farmer. However, some decision makers tend to dismiss the potential value of predictions for decision making because of uncertainty about the accuracy of the forecasts, confusion arising from forecasts coming from different sources at the same time, or cursory analyses found no potential value.

Murphy (1993) noted that forecasts must reflect our uncertainty in order to satisfy the basic maxim of forecasting—that a forecast should always correspond to a forecaster's best judgment. This means that forecasts must be expressed in probabilistic terms, because the atmosphere is not completely deterministic. In addition, the degree of uncertainty expressed in the forecast must correspond with that embodied in the preparation of the forecast.

Pfaff et al. (1999) noted that whoever has a reliable forecast first is in a position to use it to his or her advantage. To ensure that a drought EWS provides benefits to all, the communication system must be transparent—that is, the information and the process by which that information is gathered, analyzed, and disseminated needs to be open to all (Glantz, 2001). Such transparency can increase trust between potential users and the providers of the forecast information. Inter-ministerial rivalries (e.g., between agricultural ministries and meteorological services) and jurisdictional disputes must be set aside to ensure optimum use of a drought EWS.

The above description of problems in the use of climate predictions probably seems depressing. However, the adoption of a systems approach (Hammer, 2000) to drought forecasting and mitigation can help to minimize if not avoid such impediments. As Broad and Agrawala (2000) put it, for climate prediction to be useful in drought mitigation, we "must forge a partnership with society that is based on a clear understanding of social needs and a transparent presentation of its [the prediction's] own potential contribution."

IV. CLIMATE CHANGE AND DROUGHT MITIGATION

Nicholls (2004) demonstrated that record warm temperatures in Australia accompanying the 2002–2003 drought were likely the result of a continuation of the apparently inexorable warming seen since the mid 20th century. In turn, the possibility that such warming is at least partly due to the enhanced greenhouse effect and, therefore, likely to continue in the future is difficult to ignore. The record warm temperatures exacerbated the 2002 drought, by increasing evaporation and the curing of fuels for wildfires. Thus, even though the severity of the drought, as measured by rainfall deficiencies, was no lower than other droughts (e.g., in 1961 and 1994), the 2002 drought was likely more severe. Similar effects are expected across much of the globe in the future because of the enhanced greenhouse effect (IPCC, 2001), with increased summer drying and associated risk of drought and with warming likely to lead to greater extremes of drying.

What do such changes mean for the use of climate predictions for drought mitigation? First, it will be necessary to predict temperatures as well as rainfall, even in areas where, traditionally, rainfall has been the variable leading to drought hardship. Second, these temperature and rainfall forecasts will need to be synthesized into a drought forecast; this will require more sophisticated drought monitoring systems able to take into account the effect of changes in meteorological variables other than rainfall. Third, any forecast system will need to take account of the long-term climate changes (in both temperature and rainfall); it will be incorrect to assume that climate is variable but statistically stationary in the future. Finally, all the aspects will need to be communicated to users if the forecasts are to be used in the future as effectively as they might have been used before climate change.

REFERENCES

Alexander, M; Capotondi, A; Diaz, H; Hoerling, M; Huang, H; Quan, X; Sun, D; Wieckmann, K. Decadal climate and global change research. In: *NOAA 2002 Climate Diagnostics Center Science Review*, Chapter 5, pp. 71–86, *http://www.cdc.noaa.gov/review/index.html*, 2001.

Ayalew, M. What is food security and famine and hunger? *Internet Journal for African Studies* 2, *http://www.brad.ac.uk/research/ijas/ijasno2/ayalew.html,* 1997.

Broad, K; Agrawala, S. The Ethiopia food crisis—Uses and limits of climate forecasts. *Science* 289:1693–1694, 2000.

Buchanan-Smith, M. What is a famine early warning system? Can it prevent famine? *Internet Journal for African Studies* 2, *http://www.brad.ac.uk/research/ijas/ijasno2/smith.html,* 1997.

Dilley, M. Reducing vulnerability to climate variability in Southern Africa: The growing role of climate information. *Climatic Change* 45:63–73, 2000.

Drosdowsky, W. Potential predictability of winter rainfall over southern and eastern Australia using Indian Ocean sea surface temperature anomalies. *Australian Meteorological Magazine* 42:1–6, 1993.

Drosdowsky, W; Chambers, L. Near Global Sea Surface Temperature Anomalies As Predictors of Australian Seasonal Rainfall. Bureau of Meteorology Research Report No. 5, Australia, 1998.

Felts, AA; Smith, DJ. Communicating climate research to policy makers. In: HF Diaz, RS Pulwarty, eds. *Hurricanes: Climate and Socioeconomic Impacts* (pp. 234–249). Berlin: Springer-Verlag, 1997.

Fischhoff, B. What forecasts (seem to) mean. *International Journal of Forecasting* 10:387–403, 1994.

Glantz, MH. The value of a long-range weather forecast for the West African Sahel. *Bulletin of the American Meteorological Society* 58:150–158, 1977.

Glantz, MH. Eradicating famines in theory and practice: Thoughts on early warning systems. *Internet Journal for African Studies* 2, *http://www.brad.ac.uk/research/ijas/ijasno2/glantz.html,* 1997.

Glantz, MH; ed. *Once Burned, Twice Shy? Lessons Learned From the 1997–98 El Niño.* New York: United Nations University, 2001.

Hammer, GL. A general systems approach to applying seasonal climate forecasts. In: GL Hammer, N Nicholls, C Mitchell, eds. *Applications of Seasonal Climate Forecasting in Agricultural and Natural Ecosystems* (pp. 51–65). Dordrecht: Kluwer, 2000.

Hulme, M; Biot, Y; Borton, J; Buchanan-Smith, M; Davies, S; Folland, C; Nicholls, N; Seddon, D; Ward, N. Seasonal rainfall forecasting for Africa. Part II—Application and impact assessment. *International Journal of Environmental Studies* 40:103–121, 1992.

Hurrell, JW. Decadal trends in the North Atlantic Oscillation regional temperatures and precipitation. *Science* 269:676–679, 1995.

Huss-Ashmore, R. Local-level data for use as early warning indicators. *Internet Journal for African Studies* 2, *http://www.brad.ac.uk/research/ijas/ijasno2/ashmore.html,* 1997.

IFRC. *World Disasters Report. International Federation of Red Cross and Red Cresent Societies*. Dordrecht: Martinus Nijhoff, 1995.

IOCI. *Climate Variability and Change in South West Western Australia.* Perth: Indian Ocean Climate Initiative Panel, 2002.

IPCC. *Climate Change 2001. Synthesis Report. Intergovernmental Panel on Climate Change.* Cambridge: Cambridge University Press, 2001.

Kumar, A; Hoerling, P. Analysis of a conceptual model of seasonal climate variability and implications for seasonal prediction. *Bulletin of the American Meteorological Society* 81:255–264, 2000.

McHugh, M; Rogers, JC. North Atlantic Oscillation influence on precipitation variability around the southeast African convergence zone. *Journal of Climate* 14:3631–3642, 2001.

Miller, AJ; Schneider, N. Interdecadal climate regime dynamics in the North Pacific Ocean: Theories, observations and ecosystem impacts. *Progress in Oceanography* 47:355–379, 2000.

Murphy, AH. What is a good forecast? An essay on the nature of goodness in weather forecasting. *Weather and Forecasting* 8:281–293, 1993.

Murton, B. Events and context: Frameworks for the analysis of the famine process. In: HG Bohle, T Cannon, G Hugo, FN Ibrahim, eds. *Famine and Food Security in Africa and Asia: Indigenous Response and External Intervention to Avoid Hunger* (pp. 167–184). Bayreuth, Germany: Naturwissenschaftliche Gesellschaft Bayreuth, 1991.

Nicholls, N. The Southern Oscillation and Indonesian Sea surface temperature. *Monthly Weather Review* 112:424–432, 1984.

Nicholls, N. Opportunities to improve the use of seasonal climate forecasts. In: GL Hammer, N Nicholls, C Mitchell, eds. *Applications of Seasonal Climate Forecasting in Agricultural and Natural Ecosystems* (pp. 309–327). Dordrecht: Kluwer, 2000.

Nicholls, N. The changing nature of Australian droughts. *Climatic Change* (63:323–336), 2004.

Nicholls, N; Kestin, T. Communicating climate. *Climatic Change* 40:417–420, 1998.

Nigam, S; Barlow, M; Berbery, EH. Analysis links Pacific decadal variability to drought and streamflow in United States. *EOS* 80(61), 1999.

Palmer, TN; Alessandri, A; Andersen, A; Cantelaube, UP; Davey, M; Délécluse, P; Déqué, M; Díez, E; Doblas-Reyes, FJ; Feddersen, H; Graham, R; Gualdi, S; Guérémy, J-F; Hagedorn, R; Hoshen, N; Keenlyside, N; Latif, M; Lazar, A; Maisonnave, E; Marletto, V; Morse, AP; Orfila, B; Rogel, P; Terres, J-M; Thomson, MC. Development of a European multi-model ensemble system for seasonal-to-interannual prediction (DEMETER). *Bulletin of the American Meteorological Society* 85:853–872, 2004.

Pfaff, A; Broad, K; Glantz, MH. Who benefits from climate forecasts? *Nature* 397: 645–646, 1999.

Rajagopalan, B; Cook, E; Lall, U; Ray, BK. Spatiotemporal variability of ENSO and SST teleconnections to summer drought over the United States during the twentieth century. *Journal of Climate* 13:4244–4255, 2000.

Saidy, D. Early warning and response. *Internet Journal for African Studies* 2, *http://www.brad.ac.uk/research/ijas/ijasno2/saidy.html*, 1997.

Schiller, A; Godfrey, S. Indian Ocean intraseasonal variability in an ocean general circulation model. *Journal of Climate* 16:21–39, 2003.

Slingo, JM; Rowell, DP; Sperber, KR; Nortley, F. On the predictability of the interannual behaviour of the Madden-Julian Oscillation and its relationship to El Niño. Proceedings of a Conference on the TOGA Coupled Ocean-Atmosphere Response Experiment (COARE), Boulder, Colorado, 7–14 July 1998 (COARE-98), 1999.

Smith, NR. The global ocean data assimilation experiment. *Advances in Space Research* 25:1089–1098, 2000.

Stockdale, TN; Anderson, DLT; Alves, JOS; Balmaseda, MA. Global seasonal rainfall forecasts using a coupled ocean-atmosphere model. *Nature* 392:370–373, 1998.

Stone, RC; Auliciems, A. SOI phase relationships with rainfall in eastern Australia. *International Journal of Climatology* 9:1896–1909, 1992.

Troup, AJ. The Southern Oscillation. *Quarterly Journal of the Royal Meteorological Society* 91:490–506, 1965.

Walker, P. *Famine Early Warning Systems: Victims and Destitution.* London: Earthscan Publications, 1989.

Wilhite, DA. *Planning for Drought: A Process for State Government.* IDIC Technical Report Series 90-1. Lincoln, NE: International Drought Information Center, University of Nebraska–Lincoln, 1990.

Wolter, K; Timlin, MS. Monitoring ENSO in COADS with a seasonally adjusted principal component index. Proceedings of the 17th Climate Diagnostics Workshop, pp. 52–57, Norman, Oklahoma, NOAA/N MC/CAC, NSSL, Oklahoma Climatological Survey, CIMMS and the School of Meteorology, University of Oklahoma, 1993.

World Food Summit. *Plan of Action*, paragraph 1, *http://www.fao.org/docrep/003/w3613e/w3613e00.htm#PoA*, 1996.

Zeng, N; Neelin, JD; Lau, K-M; Tucker, EJ. Enhancement of interdecadal climate variability in the Sahel by vegetation interaction. *Science* 286:1537–1540, 1999.

3

Drought Monitoring: New Tools for the 21st Century

MICHAEL J. HAYES, MARK SVOBODA,
DOUGLAS LE COMTE, KELLY T. REDMOND,
AND PHIL PASTERIS

CONTENTS

I. INTRODUCTION: THE IMPORTANCE OF DROUGHT MONITORING

As the world moves into the 21st century, the stresses on available water resources will continue to grow. In the United States, increasing growth and development are already straining water supplies not only for the major metropolitan areas of the arid West, but also for areas such as Atlanta, Georgia, in the relatively humid eastern United States. Issues surrounding shared water resources across international boundaries, such as the Colorado and Rio Grande River basins between the United States and Mexico and the Great Lakes and Columbia River basins between the United States and Canada, will also continue to grow. Droughts, as a normal natural hazard in most climates, will compound these concerns. Therefore, because of serious drought impacts on water resources-related issues, planning for and responding effectively to future droughts will be critically important.

A key component to drought risk management and to breaking the "hydro-illogical cycle" (illustrated in Chapter 5) is drought monitoring. Decision makers need timely and accurate information about the development of drought conditions—in effect, an early warning system so they can anticipate the onset of drought and be prepared. They also need accurate and timely assessments of drought severity so appropriate responses can be coupled with current or anticipated drought impacts. In addition, during drought recovery, decision makers need information that can document the status of recovery and identify if and when the event is over. Drought monitoring must be a continuous process so the hazard and its impacts do not creep up on a region. Decision makers would also benefit from short- and long-term drought forecasting tools that allow them to anticipate and respond to a drought episode with greater precision.

One constraint to effective drought monitoring has always been the lack of a universally accepted definition. Scientists and decision makers must accept that the search for a single definition of drought is a hopeless exercise. Drought definitions must be specific to the region, application, or impact. Drought must be characterized by many different climate and water supply indicators. Impacts are complex and vary regionally and on temporal timescales. As described by Steinemann et al. (Chapter 4), drought monitoring indicators, ideally, should be tied directly to triggers that assist decision makers with timely and effective responses before and during drought events.

The need for improved drought monitoring is highlighted by recent widespread and severe droughts that have resulted in serious economic, social, and environmental impacts in many countries. In the United States, these droughts have fostered development of improved drought monitoring data and tools and collaborations between scientists. This chapter discusses some of these new developments as well as the current status of drought forecasting in the United States.

II. PAST EFFORTS

Steinemann et al. (Chapter 4) highlight several of the drought indicators used to monitor drought conditions. The fact that the Palmer Drought Severity Index (PDSI) gained so much attention and acceptance in the years following its development (Palmer, 1965), particularly in the United States, indicated that decision makers needed tools to monitor and respond to drought events. Before the PDSI, most drought monitoring efforts used some representation of precipitation, but these were largely applicable to specific locations and not appropriate for many applications (e.g., regional comparisons) (Heim, 2000).

The PDSI and its assortment of companion indices were quickly accepted because they considered both supply and demand, even if (in retrospect) imperfectly. Palmer had attempted to develop a drought index that included a simplified two-layer soil model and a demand component affected by temperature (Heim, 2000). The index also attempted to standardize for location and time, so that the values could be compared between different climate regimes. Historical calculations could easily be made, so comparisons through time at one spatial point were possible. The index provided a simple scale that decision makers and the public could associate with various levels of drought severity.

Unfortunately, the PDSI's many weaknesses and limitations have been identified over the years (Alley, 1984; Guttman, 1991; Guttman, 1998; Guttman et al., 1992; Hayes et al., 1999). Other indices and techniques have been developed in the United States to sidestep some of the weaknesses. The Surface Water Supply Index (SWSI) was developed to account for the snowmelt-based water resource characteristics of the western United States (Shafer and Dezman, 1982). The statistical and temporal properties of SWSI are not well characterized or understood, and the method has yet to be thoroughly critiqued in the manner of Alley (1984). Garen (1993) modified the original SWSI procedure to incorporate water supply forecasts during the winter season.

In 1993, a group of scientists at Colorado State University (McKee et al., 1993) developed a new drought index, the Standardized Precipitation Index (SPI). Extensive studies showed that the PDSI was highly correlated (typically $r > 0.90$) with precipitation at certain timescales (almost always 6–12 months), and therefore temperature added little supplementary information. Although based on precipitation alone, the SPI was designed to address many of the weaknesses associated with the PDSI and intended to provide a direct answer to the questions most commonly posed by water managers. The SPI "suite" provides information on absolute and relative precipitation deficits and excesses on a variety of timescales and on the frequency or likelihood of occurrence. The SPI can clearly show situations that are simultaneously in excess and deficit on different timescales (e.g., short wet episodes within long dry periods, or vice versa) and highlights rather than overlooks such common behavior. To date, this greatest strength of the SPI has been its least exploited.

Even with the development of the SPI and SWSI, four major limitations to drought monitoring remain:

1. *Temporal frequency of data collection.* Most changes are slow, but drought status can change appreciably in the course of a day (e.g., with heavy rain or snow), or over a few days to a week (e.g., spells of high heat,

low humidity, high winds, significant evaporation or sublimation of snowpack, sensitive phenologic stages, shallow soil moisture depletion by high evapotranspiration demand, and so forth). Thus, daily updates represent about the right frequency of new information. In many instances, data are measured and collected only on monthly, or sometimes weekly, timetables and are often unavailable for several days or weeks because of manually intensive processing methods. Concerns about the quality of near-real-time data and the quality control process involved in providing of usable recent data have also contributed to these temporal limitations.

2. *Spatial resolution.* In most cases spatial resolution has been at a coarse, regional scale. In the United States, much of the climate information has been organized by climate divisions. These divisions fail to provide the required spatial detail of drought conditions needed by decision makers, especially in the West, where topographic gradients predominate.
3. *Use of a single indicator or index to represent the diversity and complexity of drought conditions and impacts.* Decision makers need multiple indicators to understand the spatial pattern and temporal timescales within and between regions.
4. *Lack of reliable drought forecasting products.* To respond effectively, decision makers need to anticipate the development and cessation of a drought event and its progression.

III. NEW DEVELOPMENTS

During the past decade, significant progress has been made in developing new drought monitoring tools and revising some of the existing tools. Most of these developments have improved the temporal and spatial resolution of drought monitoring, providing better information to decision makers regarding specific events. Other developments include near-real-time access to data and improved information sharing, available as a result of the Internet; satellite technology; geographic information systems (GIS); and supercomputing capabilities.

A. U.S. Drought Monitor

One of the best examples of a new drought monitoring tool is the U.S. Drought Monitor (*http://drought.unl.edu/dm*). The National Drought Mitigation Center (NDMC), U.S. Department of Agriculture (USDA), and National Oceanic and Atmospheric Administration's (NOAA) Climate Prediction Center (CPC) and National Climatic Data Center (NCDC) author the weekly Drought Monitor (DM) map, which was first released in 1999. The DM is not a forecast; rather, it was designed as a comprehensive drought assessment that reflects the existing drought situation across the country. Because multiple physical conditions may be present at once and no preferred scale exists for assessing drought, the DM also incorporates and heavily weights human expertise and judgment in the assessment of associated impacts.

The DM defines four categories of drought severity based on increasing intensity (D1–D4), with a fifth category (D0) indicating an abnormally dry area (possible emerging drought conditions or an area that is recovering from drought but may still be seeing lingering impacts). The drought categories represented by this scale are moderate (D1), severe (D2), extreme (D3), and exceptional (D4).

Several characteristics of the DM product make it unique and successful. One of its strengths is that the five categories are based on a percentile approach, where D0 is approximately equal to the 30th percentile; D1, the 20th; D2, the 10th; D3, the 5th; and D4, the 2nd (Svoboda et al., 2002a).

A second key strength of the DM product is that it is based on multiple indicators. One indicator is not adequate to represent the complex characteristics of drought across a region. Therefore, it is important that a product like the DM use a variety of quantitative and qualitative indicators. The key indicators used in creating the weekly DM map include streamflow, measures of recent precipitation, drought indices, remotely sensed products, and modeled soil moisture. Many other ancillary indicators are also used, depending on the region and the season. For example, in the western United States, indicators such as snow water content, reservoir information, and water supply indices are important for evaluating the current and future availability of water. These indicators inherently incorporate the effects of hydrological lag and relationships across space and time between climate and the surface or groundwater system.

The Drought Monitor also incorporates information from approximately 150 scientists and local experts around the country. The DM seeks corroborative impact information to provide added confidence in the initial assessments gained from purely quantitative information describing the physical environment. This kind of "ground truth" is important and increases broad-based credibility of the product with users.

The Drought Monitor has performed another equally important role by focusing discussions of drought issues and, in particular, the need for additional information and products. In this regard, the DM has been an unqualified success. Building on the U.S. experience, drought experts in Canada, Mexico, and the United States developed the experimental North American Drought Monitor (NADM) product in 2002 and produce monthly reports (*http://www.ncdc.noaa.gov/oa/climate/monitoring/drought/nadm/*). The NADM represents an important step in a cooperative, multinational effort to improve monitoring and assessment of climate extremes throughout the continent (Lawrimore et al., 2002). Other nations and regions have expressed interest in a product like the DM.

Products resulting from the DM process include the CPC weekly short- and long-term products, which "blend" multiple quantitative indicators used in making the DM. The two blends attempt to identify drought severity differences resulting from shorter and longer timescales. These products, known as the Objective Blends of Drought Indicators (OBDI), are available at the CPC website (*http://www.cpc.ncep.noaa.gov/products/predictions/experimental/edb/droughtblend-access-page.html*).

B. Climate Delivery Systems

New tools and technologies have improved the capacity to monitor real-time precipitation measurements around the United States, largely resulting from the development of the Applied Climate Information System (ACIS) (Pasteris et al., 1997). The primary goals of ACIS are to integrate data from several unique networks into one transparent database maintained by the six NOAA regional climate centers and provide software tools to create a wide variety of climate-related products. A web-based interface provides access to near-real-time National Weather Service (NWS) Cooperative (COOP) Observer Network data, NCDC preliminary and historical datasets, and regional climate center network datasets. ACIS will eventually include information from a variety of federal networks (i.e., SNOTEL [SNOwpack TELemetry], SCAN [Soil Climate and Analysis Network], the Remote Automated Weather Stations network) and other regional and state Mesonet data from around the United States.

A related tool is currently being developed as part of a collaborative project between USDA's Risk Management Agency and the Department of Computer Science and Engineering, the NDMC, and the High Plains Regional Climate Center (HPRCC), all located at the University of Nebraska. This project has led to the development of the National Agricultural Decision Support System (NADSS). The NADSS website (*http://nadss.unl.edu*) contains a collection of decision support tools designed to help agricultural producers assess a variety of climate-related risks. A national interface is being developed to enable a user to generate tabular or map products for the continental United States, or for any individual state or station. Calculations for the SPI, PDSI, a newly derived self-calibrated PDSI (Wells et al., 2004), and a soil moisture model can all be generated and presented in map or table form. This operational tool is based on preliminary, quality-controlled, near-real-time data utilizing the ACIS interface.

NOAA's plan for modernizing the COOP network through automation of the existing network and development of the

National Cooperative Mesonet (a network of networks, incorporating traditional stations, airports, other networks, and other sensors, such as soil moisture probes at many locations) can potentially support improved drought monitoring (National Weather Service, 2003). This network is the backbone of the United States' long-term climate observation network, which is vital to monitoring drought. Having access to this critical and credible data resource in near real time will sustain our ability to create products that can better meet users' spatial and temporal needs.

C. Hydrological Indicators

In the mountainous western United States, snow is the dominant form of precipitation affecting streamflow, and snowmelt supplies about three quarters of the region's streamflow. This supply generally originates in rather small and elevated areas, where precipitation is concentrated and much enhanced by interactions with topography accounting for a disproportionate fraction of the snow and eventual runoff. Frozen precipitation is much more difficult to measure than liquid, and these measurements are taken in remote and climatically hostile environments that severely tax equipment, sensor technology, and human observers. These environments receive large amounts of precipitation, even in generally arid lands. This cold precipitation reduces direct loss (by evaporation and sublimation), and the subsequent modest metering of melt rates (limited largely by the rate of energy reception from the sun) ensures that a much higher percentage of winter precipitation participates in the soil recharge phenomena than is the case with summer precipitation. Much summer precipitation, especially in hotter climates and even when intense, evaporates immediately and does not contribute to subsurface recharge. In addition, the seasonal cycle of precipitation varies greatly with elevation, and mountains and valleys that adjoin each other can have very different, and often poorly correlated, precipitation and temperature histories. Thus, high-altitude precipitation needs to be measured accurately and with sufficient spatial resolution, and separately from precipitation at low altitudes (Redmond, 2003).

The Natural Resources Conservation Service (NRCS) is the key collector and provider of snowpack data through its Snow Survey and Water Supply Forecasting program. The NRCS collects snow water equivalent, precipitation, and temperature data from nearly 700 automated SNOTEL stations and snow water equivalent and snow depth data from more than 900 manually sampled sites in 12 western states and Alaska. In addition, the NRCS collects snowpack data in Vermont, New Hampshire, Pennsylvania, and Minnesota through SCAN. The data are posted to the National Water and Climate Center homepage (*http://www.wcc.nrcs.usda.gov*) when received and are updated daily.

Real-time streamflow and groundwater monitoring systems have been established for the United States by the U.S. Geological Survey (USGS). The streamflow monitoring system, called "Water Watch," is available on the USGS website (*http://water.usgs.gov/waterwatch/*). This system has been in place for several years, but new analysis tools continue to be added to the site to improve monitoring capability. In addition to the daily streamflows available, new 7-day flows are now being updated daily, with additional experimental timescales (14-, 21-, and 28-day) that may soon be available to the public as well. These timescales provide representations of the streamflow characteristics around the United States that are less susceptible to the effects of single weather events. One useful feature of the USGS streamflow displays is that the daily and *N*-day average values are expressed in terms of percentiles derived from the entire historical record.

The USGS groundwater monitoring system, the Groundwater Climate Response Network, was developed more recently (*http://groundwaterwatch.usgs.gov/*). This system also presents information in terms of percentiles, facilitating its incorporation into the DM. Both the surface and groundwater monitoring systems utilize satellite telemetry, which allows near-real-time data transmission and the usage of nationally standardized time intervals, such as calendar days.

Individual states provide important supplemental groundwater information, particularly in the eastern United States. Despite these improvements, timely groundwater monitoring around the United States remains a critical indicator that needs further enhancement. In many sections of the country it can be difficult to find locations for monitoring wells that are unaffected by deliberate or inadvertent regional withdrawals and other subsurface manipulation of aquifers.

Currently, most states in the western United States calculate the SWSI (Shafer and Dezman, 1982). The basic approach of the SWSI is to try to depict reservoirs, streamflow, snowpack, and precipitation in nondimensionalized forms. The methods used to calculate the SWSI vary by state. However, there are plans to produce a West-wide Basin Water Index (BWI). This tool will incorporate precipitation, snowpack (snow water equivalent), streamflow, reservoir storage, and other appropriate hydroclimatic elements to assist in the preparation of the DM. The regional BWI will also be an input indicator for a new water resources monitor product being considered for development for the United States on a monthly basis (Svoboda et al., 2002b).

D. Soil Moisture

Soil moisture measurements, along with groundwater monitoring, continue to be one of the biggest needs in drought monitoring (Svoboda et al., 2002a). Soil moisture is an important indicator of agricultural drought. In addition to the progress being made by implementing soil moisture probes in several statewide Mesonets, SCAN slowly continues to expand in its goal to provide a network of soil moisture monitoring sites for the United States. SCAN has approximately 100 sites in 38 states, with plans for expansion as funds become available. Soil moisture data would benefit the DM and many other products and serve as ground truth information for satellite remote sensing products. It would also benefit the initialization and verification of soil moisture models, such as the CPC soil moisture model used in the development of the DM.

E. Satellite

Satellite observations provide an important means to monitor vegetation condition dynamics over large areas by providing timely, spatially continuous information at high resolutions. Recent developments have improved satellite's drought monitoring capabilities, particularly regarding the impacts on the seasonal cycle of vegetation growth that includes green-up, maturity, senescence, and dormancy. A collaborative team of scientists from the USGS EROS, Earth Resources Observation Systems, Data Center, the NDMC, and the HPRCC is developing a prototype drought monitoring system that integrates information from climate and satellite databases using data mining techniques (Brown et al., 2002). The goal is to model the relationships between climate-based drought indicators and satellite-derived seasonal vegetation performance using the Normalized Difference Vegetation Index (NDVI) to produce a timely and spatially detailed drought monitoring product for decision makers at all levels using the Internet as the primary delivery mechanism (*http://gis-data.usgs.gov/website/Drought_Monitoring/viewer.asp*).

F. Environmental and Qualitative Indicators

Few environmental and qualitative indicators exist, and certainly very few are examined on a regional or national scale. Potential indicators within this category include tree health and mortality, water quality measures, wildlife population and intrusion indicators, and public metrics. Many of these indicators would form the basis for an improved drought monitoring system that places far greater reliance on impacts to assess drought status and to corroborate physical data, an attribute of drought monitoring that is currently almost nonexistent in the United States and in most countries. The importance of these indicators cannot be overlooked as potential drought monitoring tools.

G. Water Management Considerations

The preceding discussion has glossed over an extremely important factor: in most locations, water systems are highly manipulated. Human decisions can affect water supply and demand, so that drought status can represent a highly non-linear, and even occasionally counterintuitive or paradoxical, response to the climate drivers and their history (Redmond, 2002).

Throughout the United States, but in the western states in particular, the physical facts of climate and its extreme variation in space, and the manner in which snow is transformed into streamflow and groundwater, have led to the establishment of a highly intricate and interwoven set of legal, cultural, historical, and physical components to the system for providing water. This system developed piecemeal for about a century and a half, and although it is slowly evolving, the monitoring system, and particularly its interpretive function, must acknowledge and account for the influence of these exceedingly influential factors. One must consider the intricacies of the system of reservoirs (and the purpose of each one); the connecting rivers, canals, and tunnels; the groundwater system; the reservoir represented by snowpack; and the state of surface cover and vegetation. If water is reserved for certain purposes (e.g., spring fish migration flows), full reservoirs may be present even during drought. Management decisions may be equally influential in determining the water supply picture. There need to be entry points for such information in descriptive tools for drought.

IV. DROUGHT FORECASTING

Decision makers must have accurate drought monitoring information to respond successfully during drought events. Accurate drought forecast information and tools about future conditions are equally important. The science of drought forecasting, however, is in its infancy. To forecast drought, it is important to know something about the causes of drought. Drought is usually established by persisting high pressure that results in dryness because of subsidence of air, more sunshine and evaporation, and the deflection of precipitation-bearing storms. This is usually part of a persistent large-scale disruption in the global circulation pattern. Scientists are looking for local or distant influences that might create such atmospheric blocking patterns.

In March 2000, NOAA's CPC launched a new drought forecast tool for the United States called the Seasonal Drought Outlook (SDO) (*http://www.cpc.ncep.noaa.gov/products/expert_assessment/seasonal_drought.html*). This tool is issued monthly at the same time as the traditional long-range seasonal temperature and precipitation forecasts. The SDO attempts to anticipate the pattern and trends for drought conditions across the country 3 months in advance. The development of the SDO incorporates a mix of tools, including statistical techniques based on historical natural and "constructed" analogues, historical drought index probabilities based on the time of the year, and various dynamical and statistical precipitation and temperature outlooks spanning various time periods. The SDO forecasts have met with mixed success so far, a major stumbling block being the difficulty of forecasting rainfall, or lack thereof, during the summer.

Given the sensitivity of the atmosphere to small-scale upper-level disturbances and the difficulty of forecasting such factors more than several days in advance, as well as the inherent "noisiness" of the rainfall pattern during summer, drought forecasts will be a challenge for a long time. That's the bad news. The good news is that continuing research on the mechanisms responsible for large-scale precipitation and drought patterns should lead to slow but useful improvements in seasonal forecasts and longer term forecasts. The forecasts will never attain the accuracy currently expected for short-term temperature and rainfall forecasts, but they should be better than flipping a coin … maybe a lot better.

V. CONCLUSION

The need for drought monitoring tools for decision makers around the world is tremendous. Most of the tools discussed in this chapter have been implemented in the United States, but they have potential applications worldwide. The DM product, for example, has received a lot of attention internationally, and various regions and nations have investigated whether it is possible to develop a similar tool. A momentum is building globally for improved drought monitoring capabilities. In the United States, NOAA and the Western Governors' Association have teamed with experts from around the country to begin to develop a National Integrated Drought Information System (NIDIS) that will not only work to improve drought monitoring techniques, but also improve how this information can be used for decision support by a wide range of decision makers. Various programs within the United Nations are promoting improved drought monitoring and planning strategies around the world. These recent developments regarding drought monitoring are encouraging and may signal that decision makers are finally interested in breaking the "hydro-illogical cycle" to reduce future drought impacts.

REFERENCES

Alley WM. The Palmer Drought Severity Index: Limitations and assumptions. *Journal of Climate and Applied Meteorology* 23:100–109, 1984.

Brown JF, T Tadesse, BC Reed. Integrating satellite and climate data for US drought mapping and monitoring: first steps. Proceeding of 15th Conference on Biometeorology and Aerobiology, pp. 147–150, Kansas City, MO, 2002.

Garen DC. Revised Surface-Water Supply Index for western United States. *Journal of Water Resources Planning and Management* 119:437–454, 1993.

Guttman NB. A sensitivity analysis of the Palmer Hydrologic Drought Index. *Water Resources Bulletin* 27:797–807, 1991.

Guttman NB. Comparing the Palmer Drought Index and the Standardized Precipitation Index. *Journal of the American Water Resources Association* 34:113–121, 1998.

Guttman NB, JR Wallis, JRM Hosking. Spatial comparability of the Palmer Drought Severity Index. *Water Resources Bulletin* 28:111–119, 1992.

Hayes MJ, MD Svoboda, DA Wilhite, OV Vanyarkho. Monitoring the 1996 drought using the Standardized Precipitation Index. *Bulletin of the American Meteorological Society* 80:429–438, 1999.

Heim RR Jr. Drought indices: A review. In: DA Wilhite, ed. *Drought: A Global Assessment*, vol. 1 (pp. 159–167). London: Routledge, 2000.

Lawrimore J, RR Heim Jr, M Svoboda, V Swail, PJ Englehart. Beginning a new era of drought monitoring across North America. *Bulletin of the American Meteorological Society* 83:1191–1192, 2002.

McKee TB, NJ Doesken, J Kleist. The relationship of drought frequency and duration to time scales. Proceedings of 8th Conference on Applied Climatology, pp. 179–184, Anaheim, CA, 1993.

National Weather Service. *Coop Modernization. http://www.nws.noaa.gov/om/coop/coopmod.htm*, 2003.

Palmer WC. *Meteorological Drought*. Research Paper No. 45. Washington, D.C.: U.S. Weather Bureau, 1965.

Pasteris P, R Reinhardt, K Robbins, C Perot. UCAN: Climate information now for the next century. Proceeding of 1st Symposium on Integrated Observing System, pp. 113–116, Long Beach, CA, 1997.

Redmond KT. The depiction of drought: A commentary. *Bulletin of the American Meteorological Society* 83:1143–1147, 2002.

Redmond KT. Climate variability in the West: Complex spatial structure associated with topography, and observational issues. In: WM Lewis Jr, ed. *Water and Climate in the Western United States* (pp. 29–48). Boulder, CO: University Press of Colorado, 2003.

Shafer BA, LE Dezman. Development of a Surface Water Supply Index (SWSI) to assess the severity of drought conditions in snowpack runoff areas. Proceedings of the Western Snow Conference, pp. 164–175, Fort Collins, CO, 1982.

Svoboda M, D Le Comte, M Hayes, R Heim, K Gleason, J Angel, B Rippey, R Tinker, M Palecki, D Stooksbury, D Miskus, S Stevens. The Drought Monitor. *Bulletin of the American Meteorological Society* 83:1181–1189, 2002a.

Svoboda MD, M Hayes, R Tinker. Three years and counting: What's new with the Drought Monitor? Proceedings of the 13th Conference on Applied Climatology, pp. 223–226, Portland, OR, 2002b.

Wells N, S Goddard, M Hayes. A self-calibrating Palmer Drought Severity Index. *Journal of Climate*, 17:2335–2351, 2004.

4

Drought Indicators and Triggers

ANNE C. STEINEMANN, MICHAEL J. HAYES,
AND LUIZ F.N. CAVALCANTI

CONTENTS

I. OVERVIEW OF INDICATORS AND TRIGGERS

Drought indicators are variables that describe the magnitude, duration, severity, and spatial extent of drought. Typical indicators are based on meteorological and hydrological variables, such as precipitation, streamflows, soil moisture, reservoir storage, and groundwater levels. Several indicators can be synthesized into a single indicator on a quantitative scale, often called a drought index. Although drought indices can provide ease of implementation, the scientific and operational meaning of an index value may raise questions, such as how each indicator is combined and weighted in the index and how an arbitrary index value relates to geophysical and statistical characteristics of drought.

Drought triggers are threshold values of an indicator that distinguish a drought level and determine when management actions should begin and end. Triggers ideally specify the indicator value, the time period, the spatial scale, the drought level, and whether conditions are progressing or receding. Drought levels (or phases or stages) are categories of drought, with nomenclature such as *mild*, *moderate*, *severe*, and *extreme*, or *stage 1*, *stage 2*, and *stage 3*.

Drought indicators and triggers are important for several reasons: to detect and monitor drought conditions; to determine the timing and level of drought responses; and to characterize and compare drought events. Operationally, they form the linchpin of a drought management plan, tying together levels of drought severity with drought responses.

Of the more than 150 published definitions of drought (Wilhite and Glantz, 1985), one theme emerges: drought is a condition of insufficient water to meet needs (Redmond, 2002). Water needs and water supplies differ depending on context, and thus the characterization of drought can require different indicators and quantifications for triggers. The most common relate to meteorological and hydrological water availability and uses (Byun and Wilhite, 1999; Heim, 2002; Wilhite and Glantz, 1985).

Meteorological drought indicators are associated with climatological variables such as precipitation, temperature, and evapotranspiration. Meteorological indices include the Palmer Drought Severity Index (PDSI) (Palmer, 1965), deciles (Gibbs and Maher, 1967), and the Standardized Precipitation Index (SPI) (McKee et al., 1993). Precipitation is a widely used and useful indicator; it can directly measure water supplies, it influences hydrological indicators, and it can reflect drought impacts over different time periods and sectors. Yet meteorological indicators, such as precipitation, can pose analytic challenges because of temporal and spatial variability, lack of data, and insufficient observation stations. The PDSI has been one of the most widely used indices in the United States, even though the SPI has advantages of statistical consistency and the ability to reflect both short- and long-term drought impacts (Guttman, 1998; Hayes et al., 1999). An evaluation of common indicators, according to six criteria of performance, indicates strengths of the SPI and deciles over the PDSI (Keyantash and Dracup, 2002).

Hydrological drought indicators relate to water system variables such as groundwater levels, streamflows, reservoir storage, soil moisture, and snowpack. Hydrological indices include the Surface Water Supply Index (SWSI) (Shafer and Dezman, 1982) and the Palmer Hydrological Drought Index (PHDI) (Karl, 1986). These indicators reflect that hydrological drought usually is slow to develop and persists longer than meteorological drought. For instance, groundwater is usually a later indication of drought conditions and a more conservative indicator for recovering from drought, yet its usefulness may be limited by poor understanding of subsurface conditions and anthropogenic influences. Streamflows can integrate other indicators, such as soil moisture, groundwater, and precipitation, yet can also be heavily influenced by anthropogenic factors, such as development and diversions. Reservoir levels and reservoir storage are easy to measure, yet operating rule curves may complicate assessments of drought conditions. The PHDI was designed to reflect longer term hydrological impacts, and the SWSI was developed to address some of the limitations of the Palmer indices by incorporating water supply information. Both of these indices, however, can be difficult to interpret directly and compare consistently. The various strengths and limitations of common indices will now be examined in greater detail.

A. Precipitation

Precipitation, as a variable, can be transformed into several types of indices:

1. *Percent of normal* can analyze a single region or a single season, yet it is easily misunderstood and gives different values depending on the location and time period. Further, mean precipitation (the average amount) usually differs from median precipitation (the amount exceeded 50% of the time) because precipitation tends to be skewed rather than normally distributed. For a positively skewed precipitation distribution, the median is less than the mean, so below-normal (below-average) precipitation is more likely than above-normal precipitation. For instance, in Melbourne, Australia, median precipitation for February is 32.4 mm, but this is only 68.6% of "normal" when compared to the mean (47.2 mm) (AU-CBM, 2003). Using percent of normal can make it difficult to link a value of a departure with a specific impact occurring as a result of the departure, and thus to design appropriate drought mitigation and response measures (Willeke et al., 1994).
2. *Deciles* (Gibbs and Maher, 1967) can address limitations of the percent of normal approach. The long-term precipitation record is divided into tenths of percentiles, called deciles: the lowest 20% is much below normal; next lowest 20% is below normal; middle 20% is near normal; next highest 20% is above normal; and highest 20% is much above normal. The deciles method was selected over the PDSI for the Australian Drought Watch System for simplicity, con-

sistency, and understandability (Smith et al., 1993). One challenge, though, is that a long climatological record with consistent observation stations is needed to calculate the deciles accurately. Also, deciles can be difficult to apply if officials and the public are not familiar with the system.

3. *Standardized Precipitation Index (SPI)*, developed by McKee et al. (1993), quantifies precipitation deficit for multiple timescales, such as for 3-, 6-, 9-, and 12-month prior periods, relative to those same months historically. These different timescales are designed to reflect the impacts of precipitation deficits on different water resources. For instance, soil moisture conditions respond to precipitation anomalies on a relatively short scale, whereas groundwater, streamflow, and reservoir storage reflect longer term precipitation anomalies.

The SPI relies on a long-term precipitation record, typically at least 30 years, for a desired region, such as a climate division. This record is fitted to a probability distribution, such as the gamma distribution or Pearson III, so that a percentile on the fitted distribution corresponds to the same percentile on a gaussian distribution (Panofsky and Brier, 1958). That percentile is then associated with a Z score for the standard gaussian distribution, and the Z score is the value of the SPI.

The categories of the SPI, according to McKee et al. (1993), are as follows:

SPI Values	Drought Category	Cumulative Frequency
0 to –0.99	Mild drought	16–50%
–1.00 to –1.49	Moderate drought	6.8–15.9%
–1.50 to –1.99	Severe drought	2.3–6.7%
–2.00 or less	Extreme drought	<2.3%

One advantage of the SPI is that it is standardized, so its values represent the same probabilities of occurrence, regardless of time period, location, and climate. A disadvantage is that the SPI values may not be intuitive to decision makers. Also, equal categorical intervals have differing probabilities of occurrence. For instance, the probability differential between an SPI of –1.0 and –1.5 is 9.1% (moderate drought); between an SPI of –1.5 and –2.0, the probability differential is 4.4% (severe drought), even though both represent an index differential of 0.5.

B. Palmer Drought Severity Index (PDSI) and Palmer Hydrologic Drought Index (PHDI)

The PDSI, based on the Palmer Drought Model (Palmer, 1965), has been one of the most commonly used drought indicators in the United States. One reason for its popularity is that its development in 1965 preceded other indices and resulted in its widespread use and wide-ranging application. The PDSI is derived from a moisture balance model, using historic records of precipitation, temperature, and the local available water capacity of the soil. The PHDI uses a modification of the PDSI to assess longer term moisture anomalies that affect streamflow, groundwater, and water storage. A primary difference between the PDSI and the PHDI is in the calculation of drought termination, using a ratio of moisture received to moisture required to definitely terminate a drought. With the PDSI, a drought ends when the ratio exceeds 0%, if it remains greater than 0% until reaching 100%. With the PHDI, a drought does not end until the ratio reaches 100% (Karl, 1986; Karl et al., 1987).

The PDSI and PHDI are calculated for climate divisions, typically on a monthly basis, with cumulative frequencies representing all months and all climate divisions (Karl, 1986):

PDSI/PHDI	Drought Category	Cumulative Frequency (approx.)
0.00 to –1.49	Near normal	28–50%
–1.50 to –2.99	Mild to moderate drought	11–27%
–3.00 to –3.99	Severe drought	5–10%
–4.00 or less	Extreme drought	4%

These PDSI/PHDI values, however, are not spatially and temporally invariant. Cumulative frequencies vary, depending on the region and time period under consideration (Guttman et al., 1992; Karl et al., 1987; Nkemdirim and Weber, 1999; Soulé, 1992). For example, the category of "extreme drought," with the overall frequency of 4%, varies in frequency from less than 1% in January in the Pacific Northwest to more than 10% in July in the Midwest (Guttman et al., 1992; Karl et al., 1987). As another example, the probability of "extreme drought" in Virginia varies as follows: Climate Division 1 for January, 4.17%; Climate Division 1 for July, 2.08%; Climate Division 6 for January, 3.21%; Climate Division 6 for July, 1.04% (Lohani et al., 1998).

As regional drought indices, the PDSI and PHDI permit comparisons of drought events over relatively large areas. The Palmer indices also offer a long-term historic record, going back more than 100 years. Yet the Palmer indices and their water balance model have several limitations (Alley, 1984; Guttman et al., 1992; Karl, 1986; Karl and Knight, 1985). The Palmer indices are not particularly suitable for droughts associated with water management systems, because they exclude water storage, snowfall, and other supplies. They also do not consider human impacts on the water balance, such as irrigation. The values for determining the severity of the drought, and the beginning and end of a drought, were arbitrarily selected based on Palmer's studies of central Iowa and Kansas. The water balance model has been critiqued on several grounds; for instance, soil moisture capacities of the two soil layers are independent of changes in vegetation. The methodology used to normalize the values is only weakly justified on a physical or statistical basis. For instance, for climatic regions with a large interannual variation of precipitation, the statistical measure of normal is less meaningful than other measures, such as the range, median, or mode of the precipitation distribution (Wilhite and Glantz, 1985). The indices are based on departures from climate normals, with no consideration of precipitation variability, so they tend not to perform well in regions with extreme variability in rainfall or runoff (Smith et al., 1993). Although the Palmer indices are widely applied within the United States, they have little acceptance elsewhere (Kogan, 1995).

C. Surface Water Supply Index

The SWSI, pronounced *swazee*, was developed by Shafer and Dezman (1982) to address limitations of the Palmer indices and incorporate water supply data, such as snow accumulation and melt, which are important in the western United States. The index is based on four components: snowpack, streamflow, precipitation, and reservoir storage. Monthly data for each component are analyzed according to probabilities of occurrence, combined into an overall index, and weighted according to their relative contributions to surface water in the basin. A modified SWSI (Garen, 1993) provides stronger statistical foundations to the index, with drought categories and cumulative frequencies as follows:

SWSI	Drought Category	Cumulative Frequency (approx.)
–2.00 to 0.00	Mild drought	26–50%
–3.00 to –2.00	Moderate drought	14–26%
–4.00 to –3.00	Severe drought	2–14%
Below –4.00	Extreme drought	<2%

An advantage of the SWSI is that it represents water supply conditions unique to each hydrological area, such as regions heavily influenced by snowpack. Limitations are that changing data sources or water supply sources require that the entire index be recalculated to account for changes in the frequency distributions and the weights of each component. For instance, discontinuing any station means that new stations need to be added to the system and new frequency distributions need to be determined for that component. Thus, a homogeneous time series of the index is difficult to maintain. If extreme events are beyond the historical time series, the index will also need to be recalculated. Further, because the index is unique to each basin, comparisons among basins or regions are limited (Doesken et al., 1991).

The following sections provide guidance in the development, implementation, and evaluation of these common indicators. Although the purpose of this chapter is not to review all possible indicators and triggers, these key examples will nonetheless illustrate important and more general concepts. For additional details on specific indicators and definitions, see, for example, Dracup et al. (1980), Fisher and Palmer (1997), WMO (1992), and Heim (2002).

II. MULTIPLE INDICATORS AND TRIGGERS: CHALLENGES AND SOLUTIONS

A. Typical Problems with Indicators and Triggers

Because drought can be characterized in many different ways, and because single indicators often prove inadequate for decision makers, multiple indicators and triggers can be useful. Challenges arise in trying to combine multiple variables and values in a drought management plan. Indicator scales may be incomparable, and trigger values may be statistically inconsistent.

Comparison of the three index scales above illustrates common problems with indicators and triggers in drought plans. These problems exist not only for values of indices (e.g., SPI, PDSI/PHDI, SWSI), but also for values of indicators (e.g., total monthly precipitation, average monthly streamflow, average monthly reservoir levels), for several reasons:

First, drought categories (levels) are inconsistent in terms of cumulative frequency. For instance, "severe drought" occurs 4.4% of the time for the SPI, 5% for the PDSI/PHDI, and 12% for the SWSI. Second, index values are difficult to interpret directly (What does a –1.5 index value mean?) and imply different probabilities of occurrence for different indicators. A value of –1.5 represents a cumulative probability of 6.7% for the SPI, approximately 27% for the PDSI/PHDI, and 32% for the SWSI. Third, as we saw earlier, the values of the indicator can vary, in terms of frequencies, depending on time and location (with the exception of the SPI). Finally, because of these inconsistencies, trying to use more than one indicator in operational drought management can cause confusion and impede effective and timely drought response.

An evaluation of state drought plans in the United States reveals wide variation in quality concerning indicators and triggers. The four plan types characterized represent incremental degrees of detail:

1. The plan mentions indicators, but without details on how these indicators are measured or used. For instance, an indicator of "precipitation" is mentioned, but not whether precipitation is measured by the SPI, deciles, or another approach. Also lacking are triggers and drought plan levels.
2. The plan provides some guidance on indicators, but without information on trigger values and corresponding drought levels. For instance, a plan may say that "streamflows" are monitored by the "monthly mean values," but the values associated with drought levels and responses are not specified.
3. The plan provides indicators and triggers, typically raw values of the indicators, which often lack statistical consistency. For instance, the plan may use the SPI-6, PHDI, and streamflows, but these indicator values have different probabilities of triggering each drought plan level. Thus, some indicators influence triggering more than others. Even if the plan specifies how the triggers may be combined, that combination method may also be statistically inconsistent.
4. The plan contains details on indicators and trigger values, plus triggers and associated drought levels are statistically comparable. One way to accomplish this is through a percentile-based approach, which is described in the next section.

B. Percentiles for Drought Indicators and Triggers

A solution for using multiple and often statistically inconsistent indicators in a drought management plan is to transform all indicators, triggers, and drought levels to a scale based on percentiles. Then trigger values are associated with the percentiles defining the drought levels. The usefulness of this approach becomes apparent when trying to compare, combine, and choose among drought indicators and trigger values. It offers a consistent and equitable basis for evaluation, ease of interpretation, and application to water management decisions, such as by relating triggers to familiar concepts as return periods and probabilities of occurrence.

Indicators and indices can be transformed to percentiles by fitting a distribution to the data (such as a gamma distribution or Pearson III for precipitation) or by developing an empirical cumulative distribution function (ECDF) using ranking algorithms, plotting positions, or other cumulative probability estimators (Harter, 1994). The drought plan triggers are then based on percentiles instead of raw indicator or index values.

For instance, suppose one uses the PDSI as an indicator in a drought plan. Rather than specify a single PDSI value (such as –1.5) for triggering a drought level (such as moderate drought) for all locations and time periods, instead specify either a percentile (such as 0.20) or a PDSI value associated with that percentile for each location and time period (e.g., 0.20 associated with a PDSI value of –1.3 for January for Climate Division 1, a value of –1.2 for February for Climate Division 1, and so forth for each month for each climate division).

To do this, stratify the long-term record of PDSI data by location (such as climate division) and time period (such as month) or in subsets of data that approximate stationarity. Then develop an ECDF for each stratified dataset, using a method such as: $p(x_i) = (i)/(n + 1)$, where $p(x_i)$ is the cumulative probability estimator; x_i is the value of the drought indicator; i is the rank of the order statistic x_i, where $i = 1, \ldots, n$ (in order from smallest to largest values); and n is the number of data values. Next, select the desired percentile for triggering a certain level of a drought plan (such as the 20th percentile for moderate drought). Using each ECDF, extract or extrapolate the PDSI value associated with that percentile. A similar approach can be used for other indicators, whether based on a theoretical or empirical distribution, such that each trigger value associated with each drought level is statistically consistent (Steinemann, 2003).

C. Example: The U.S. Drought Monitor

One example of a product developed from multiple indicators is the weekly U.S. Drought Monitor product (see Chapter 3). This product, originally released in August 1999, was developed to provide a weekly assessment of drought conditions across the United States on a general scale. What makes the Drought Monitor unique is that it incorporates a variety of quantitative indicators and is adjusted based on qualitative information from a network of local experts around the country. The quantitative indicators include the Palmer and Standardized Precipitation Indices, streamflow information, a soil moisture model, precipitation totals for various time periods, and a vegetation index derived from satellite data. Although some of this information is available in percentiles, the map derives from a subjective combination of this information and the qualitative indicators.

An advantage of the Drought Monitor is that the map provides a "big picture" assessment of drought conditions across the country for the public, media, policy makers, and others interested in a relatively simple representation of the overall drought situation. It also recognizes that because of the complexity of drought conditions and impacts, it is important to make adjustments to the drought depiction based on the qualitative information. The network of local experts provides a crucial accountability process to make sure the Drought Monitor map is representing drought conditions at this larger scale. The Drought Monitor, however, is not meant to capture local drought conditions, and this is a major limitation. It should not be used for making decisions at smaller resolutions, representing counties, for example.

III. DEVELOPING AND EVALUATING INDICATORS AND TRIGGERS

From a systematic review of state and local drought plans and interviews with water officials, we have developed a set of considerations and criteria for indicators and triggers.

A. Considerations for Drought Indicators and Triggers

1. *Suitability for drought types of concern*. An indicator needs to reflect the type of drought of concern, including aspects of water demands, water supplies, drought vulnerabilities, and potential impacts. Because drought depends on numerous factors, no single indicator is likely to cover all types of drought. In choosing indicators, a first consideration is that they should make sense for the context. For instance, the Palmer indices may not be appropriate as sole indicators for managed water systems because they do not incorporate reservoir storage. Reservoir storage, on the other hand, may not be appropriate as a sole indicator for agricultural areas that use only groundwater for irrigation.
2. *Data availability and consistency.* The performance of an indicator depends on the availability and quality of the data. Many indicators may be conceptually attractive, but are difficult, costly, unreliable, or impractical to generate, so they may not be appropriate for use. When choosing an indicator, consider the following questions: Are the data readily available? Is the indicator straightforward to calculate? Are the data trustworthy? Will the analytic expense justify the decision-making value? Does the value of the indicator vary, depending on the source of data or method of calculation? Is there a consistent long-term record, and will the data be consistently generated in the future? Many drought plans use indicators based on data that are already collected, subjected to

quality control, and consistently reported, such as by a government agency.

3. *Clarity and validity.* Indicators and triggers need to be readily understood and scientifically sound, so that drought decisions can be made and defended on the basis of them. In addition, they should be tested before a drought and evaluated after a drought to see how well they performed. A pre-drought assessment could involve generating historic sequences of triggers and comparing them to human assessments of the drought triggers that should have been invoked during that time. Another approach is to conduct virtual drought exercises with stakeholders and decision makers, using different sets of triggers and comparing management responses. A post-drought evaluation could involve a similar process of comparing actual triggers invoked to triggers that would have provided the greatest decision-making value.
4. *Temporal and spatial sensitivity.* Indicators and triggers need to consider both temporal and spatial variability, because indicator levels that imply drought conditions for one time period or one region could imply wet or normal conditions for another time period or region. For instance, "monthly total precipitation of 3 inches" could imply dry conditions in early spring but wet conditions in late summer, or imply dry conditions for a mountainous area of a state but wet conditions for a desert area of the same state.
5. *Temporal and spatial specificity.* Indicators and triggers should specify the temporal and spatial scale of analysis. Indicators need to be associated with a specific period of calculation. For instance, the SPI-6 calculates the precipitation anomaly for a prior 6-month period relative to that same 6-month period historically. Triggers also need to be associated with a time period for determining drought levels and responses. For example, "SPI-6 below –1.5 for two consecutive months would invoke Level 2 drought." In this case, the indicator's time period would be the

six prior months and the trigger's time period would be two consecutive months. Indicators also should define the scale of analysis, such as a climate division or hydrological basin for precipitation, soil moisture, and snowpack. For other indicators, such as groundwater, reservoir levels and storage, and streamflows, the spatial scale may be implicitly defined by the selection of a specific well, reservoir, or gauging station. Indicators should, nevertheless, specify the spatial scale of drought that they seek to represent, such as a set of streamflows to represent drought within a certain river basin. Triggers need to define the spatial scale of implementation of drought responses, such as the use of three groundwater wells to trigger drought responses within an entire climate division or county. Even a trigger such as a reservoir level does not necessarily imply that the spatial scale of response is that reservoir, but instead could trigger responses, such as water use restrictions, for an entire state.

6. *Drought progression and recession*. At each level of a drought plan, indicators and triggers should be defined for getting into and out of a drought,. Even though many drought plans assume that the progressing triggers can be reversed to function as the receding triggers, this approach may not be desirable from a drought management perspective. Different management goals may exist for going into a drought versus coming out of a drought. For instance, it may be important to implement water use restrictions as soon as drought conditions start developing, but to be more conservative and wait to lift restrictions when drought conditions appear to be recovering. Trigger examples would be to invoke a drought level after two consecutive months in a certain or more severe level, but to wait to revoke drought restrictions until after four consecutive months in a certain or less severe level.

7. *Statistical consistency.* Triggers need to be statistically consistent with drought levels and other triggers in a drought plan. As we saw earlier, the probabilities of occurrence of the Palmer index were not consistent among drought levels and varied according to time and location. Moreover, the index scales of the PDSI/PHDI, SPI, and SWSI were not consistent with each other. For instance, the value of –1.5 had different cumulative frequencies for each index. From the perspective of a decision maker, choosing drought indicators may be difficult, but that difficulty will be compounded if indicator scales and trigger values cannot be validly compared and combined in drought decision making.
8. *Linked with drought management goals.* Indicators need to be linked with drought management and impact reduction goals, and trigger levels should be set to invoke responses at times and stages that are consistent with these goals. For instance, triggers can be set so that a certain percentile will invoke responses that will produce a desired percentage reduction in water use. One should also consider drought indicator performance; for instance, the degree of responsiveness or persistence desired in an indicator. Some water managers may prefer an indicator that responds quickly to short-term anomalies, such as the SPI-3, so they can take early action to reduce drought impacts, whereas other water managers may prefer an indicator with greater stability and persistence, such as the SPI-12, to avoid frequent invoking and revoking of drought responses. Intermediate indicators, such as the SPI-6, can provide elements of both.
9. *Explicit combination methods.* Drought plans often rely on multiple indicators. But a question arises: How are multiple indicators considered or combined to determine a final drought level? Multiple triggers may suggest different drought levels, so methods need to be specified for combining triggers and deter-

mining the final level. These can include quantitative methods and criteria such as: the most severe of the indicators, at least one of the indicators, or a majority of indicators. These can also include qualitative methods, such as convening a drought committee to determine when to implement responses. Whether quantitative or qualitative, the methods for calculating indicators and the process for combining opinions or weighting individual data for an overall indicator should be specified.

10. *Quantitative and qualitative indicators.* Indicators can be based on quantitative data and qualitative assessments, or both. Although many drought plans center on quantitative indicators, the importance of qualitative expertise should not be overlooked. A human expert is able to consider and synthesize numerous indicators, applying years of experience and expertise to assess drought conditions. Perhaps even more important is the recognition that indicators and triggers are meant to help decision makers, not replace them. A drought plan is only one source of information, and other considerations will likely be important for decision making.

B. Checklist for Indicators and Triggers in a Drought Plan

In addition to the above criteria and considerations, we provide a checklist below. Note that these pertain only to the indicators and triggers portion of a drought plan. Many other drought plan components are important, such as communication and coordination among agencies responsible for monitoring the indicators and implementing the responses if they are triggered (see Chapter 5). Nonetheless, this list offers a straightforward set of metrics to check:

1. Indicator specification and consistency
   a. Indicator definitions: Is each indicator defined and specified?

   b. Indicator calculation method: Is a calculation method provided for each indicator?
   c. Spatial scale definition: Is the spatial scale for monitoring and analyzing the indicator defined (such as a particular climate division for the SPI-6)?
   d. Spatial sensitivity: Is the indicator sensitive to spatial variability, such as wetter regions in the mountains and drier regions in the desert?
   e. Temporal scale definition: Is the temporal scale for monitoring and analyzing the indicator defined (such as a prior 6-month period for the SPI-6)?
   f. Temporal sensitivity: Is the indicator sensitive to temporal variability, such as wetter months in the early spring and drier months in the late summer?
2. Trigger and drought level specification and consistency
   a. Definition of drought levels: Are explicit drought levels defined, such as "level/stage/phase 0, 1, 2, and 3" or "normal, moderate, severe, extreme"?
   b. Definition of triggering thresholds for each indicator: Are quantitative indicator thresholds (i.e., triggers) defined for each drought level?
   c. Spatial scale of triggers: Do triggers specify the spatial scale for implementation? Consider the trigger, "SPI-6 less than –1.5 for two consecutive months within Climate Division 1 will invoke Level 2 drought responses for counties X and Y." Here, the spatial scale for the trigger would be Counties X and Y, even though the spatial scale for the indicator is Climate Division 1.
   d. Temporal scale of triggers: Do triggers specify the temporal scale for implementation? Consider the trigger, "SPI-6 less than –1.5 for two consecutive months within Climate Division 1 will invoke Level 2 drought responses for Counties X and Y."

Here, the temporal scale for the trigger would be two consecutive months, even though the temporal scale for the indicator is the prior 6 months.

e. Statistical consistency among indicators, triggers thresholds, and drought levels: Are the triggers (one or more) statistically consistent with each other and with drought levels?
f. Explicit triggers for "drought progressing": Are indicators and triggers defined for moving from a less severe drought level to a more severe drought level?
g. Explicit triggers for "drought receding": Are indicators and triggers defined for moving from a more severe drought level to a less severe drought level?
h. Explicit method to combine indicators: Are objective or subjective methods for using and combining multiple indicators and triggers specified?

IV. CONCLUSION

Indicators and triggers are essential to drought preparation and response, yet they often lack needed attention in drought plans and planning processes. This chapter examined the most common indicators and triggers, and provided guidance for their development and evaluation. What makes a "good" indicator or trigger depends not only on its scientific merits, but also on its value to decision makers.

ACKNOWLEDGMENTS

This work received support from the National Science Foundation under grant CMS 9874391. Any opinions, findings, or conclusions are those of the authors and do not necessarily reflect the views of the National Science Foundation.

REFERENCES

Alley, WM. The Palmer Drought Severity Index: Limitations and assumptions. *Journal of Climate and Applied Meteorology* 23:1100–1109, 1984.

AU-CBM (Australia Commonwealth Bureau of Meteorology). Climate Averages for Australian Sites. *http://www.bom.gov.au/climate/averages/tables/cw_086071.shtml*, website accessed November 22, 2003.

Byun, H; Wilhite, DA. Objective quantification of drought severity and duration. *Journal of Climate* 12:2747–2756, 1999.

Doesken, NJ; McKee, TB; Kleist, J. Development of a surface water supply index for the western United States. Climatology Report Number 91–3, Colorado State University, Fort Collins, CO, 1991.

Dracup, JA; Lee, KS; Paulson, EG Jr. On the definition of droughts. *Water Resources Research* 16(2):297–302, 1980.

Fisher, S; Palmer, RN. Managing water supplies during drought: Triggers for operational responses. *Water Resources Update* 3(108):14–31, 1997.

Garen, DC. Revised Surface-Water Supply Index for western United States. *Journal of Water Resources Planning and Management* 119(4):437–454, 1993.

Gibbs, WJ; Maher, JV. Rainfall deciles as drought indicators. Bureau of Meteorology Bulletin No. 48, Commonwealth of Australia, Melbourne, 1967.

Guttman, NB. Comparing the Palmer Drought Severity Index and the Standardized Precipitation Index. *Journal of the American Water Resources Association* 34(1):113–121, 1998.

Guttman, NB; Wallis, JR; Hosking, JRM. Spatial comparability of the Palmer Drought Severity Index. *Water Resources Bulletin* 28:1111–1119, 1992.

Harter, HL. Another look at plotting positions. *Communications in Statistics—Theory and Methods* 13:1613–1633, 1994.

Hayes, MJ; Svoboda, M; Wilhite, DA; Vanyarkho, O. Monitoring the 1996 drought using the SPI. *Bulletin of the American Meteorological Society* 80:429–438, 1999.

Heim, RR Jr. A review of twentieth-century drought indices used in the United States. *Bulletin of the American Meteorological Society* 83(8):1149-1165, 2002.

Karl, TR. The sensitivity of the Palmer Drought Severity Index and Palmer's Z-Index to their calibration coefficients including potential evapotranspiration. *Journal of Climate and Applied Meteorology* 25:77–86, 1986.

Karl, TR; Knight, RW. Atlas of monthly Palmer Hydrological Drought Indices (1931–1983) for the contiguous United States. Historical Climatology Series 3–7, National Climatic Data Center, Asheville, NC, 1985.

Karl, TR; Quinlan, F; Ezell, DZ. Drought termination and amelioration: Its climatological probability. *Journal of Climate and Applied Meteorology* 26:1198–1209, 1987.

Keyantash, J; Dracup, JA. The quantification of drought: An evaluation of drought indices. *Bulletin of the American Meteorological Society* 83(8):1167–1180, 2002.

Kogan, FN. Droughts of the late 1980s in the United States as derived from NOAA polar-orbiting satellite data. *Bulletin of the American Meteorological Society* 76(5):655–668, 1995.

Lohani, VK; Loganathan, GV; Mostaghimi, S. Long-term analysis and short-term forecasting of dry spells by Palmer Drought Severity Index. *Nordic Hydrology* 29(1):21–40, 1998.

McKee, TB; Doesken, NJ; Kleist, J. The relationship of drought frequency and duration to time scale. Preprints, Eighth Conference on Applied Climatology (pp. 179–184), Anaheim, CA, 1993.

Nkemdirim, L; Weber, L. Comparison between the droughts of the 1930s and the 1980s in the Southern Prairies of Canada. *Journal of Climate* 12:2434–2450, 1999.

Palmer, WC. Meteorological drought. Research Paper No. 45, U.S. Department of Commerce Weather Bureau, Washington, D.C., 1965.

Panofsky, HA; Brier, GW. *Some Applications of Statistics to Meteorology.* University Park, PA: Pennsylvania State University, 1958.

Redmond, K. The depiction of drought—A commentary. *Bulletin of the American Meteorological Society* 83(8):1143–1147, 2002.

Shafer, BA; Dezman, LE. Development of a Surface Water Supply Index (SWSI) to assess the severity of drought conditions in snowpack runoff areas. Proceedings of the Western Snow Conference (pp. 164–175), Fort Collins, CO, 1982.

Smith, DI; Hutchinson, MF; McArthur, RJ. Australian climatic and agricultural drought: Payments and policy. *Drought Network News* 5(3):11–12, 1993.

Soulé, PT. Spatial patterns of drought frequency and duration in the contiguous USA based on multiple drought definitions. *International Journal of Climatology* 12:11–24, 1992.

Steinemann, A. Drought indicators and triggers: A stochastic approach to evaluation. *Journal of the American Water Resources Association* 39(5):1217–1234, 2003.

Wilhite, DA; Glantz, MH. Understanding the drought phenomenon: The role of definitions. *Water International* 10:111–120, 1985.

Willeke, G; Hosking, JRM; Wallis, JR; Guttman, NB. The National Drought Atlas. Institute for Water Resources Report 94–NDS–4, U.S. Army Corps of Engineers, 1994.

WMO (World Meteorological Organization). *International Meteorologic Vocabulary*, 2nd ed. Geneva, Switzerland: WMO, No. 192, 1992.

5

Drought Preparedness Planning: Building Institutional Capacity

DONALD A. WILHITE, MICHAEL J. HAYES, AND CODY L. KNUTSON

CONTENTS

I. INTRODUCTION

Past attempts to manage drought and its impacts through a reactive, crisis management approach have been ineffective, poorly coordinated, and untimely, as illustrated by the hydro-illogical cycle in Figure 1. The crisis management approach has been followed in both developed and developing countries. Because of the ineffectiveness of this approach, greater interest has evolved in recent years in the adoption of a more proactive risk-based management approach in some countries (see Chapter 6). Other countries are striving to obtain a higher level of preparedness through development of national action programs that are part of the United Nations Convention to Combat Desertification (UNCCD) or as part of separate national initiatives. In part, these actions directly result from the occurrence of recent severe drought episodes that have persisted for several consecutive years or frequent episodes that have occurred in succession with short respites for recovery between events. Global warming, with its threat of an

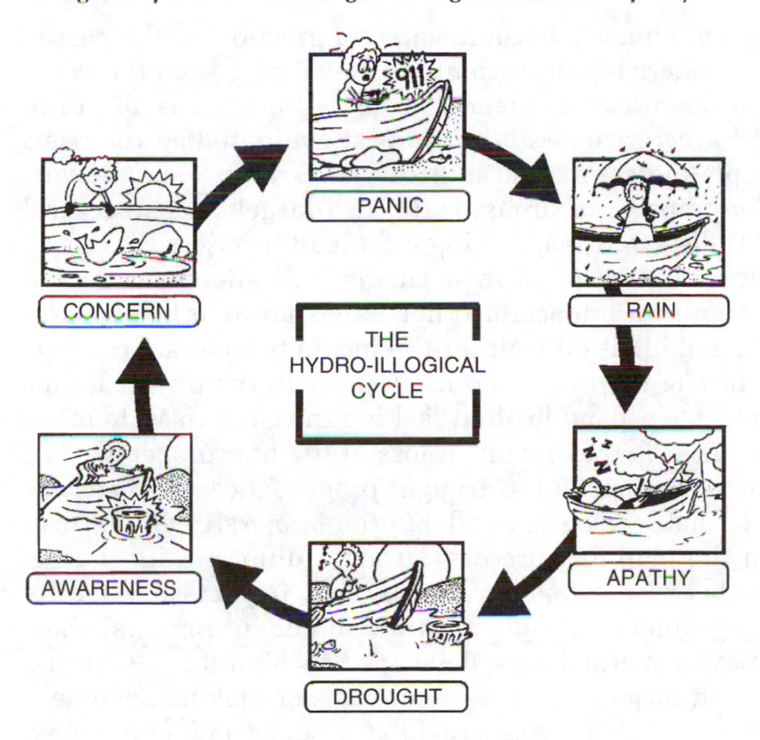

Figure 1 Hydro-illogical cycle. (*Source*: National Drought Mitigation Center, University of Nebraska, Lincoln, Nebraska, USA.)

increased frequency of drought events in the future, has also caused greater anxiety about the absence of preparation for drought, which is a normal part of climate. Other factors that have contributed to this trend toward improved drought preparedness and policy development are spiraling costs or impacts associated with drought, complexity of impacts on sectors well beyond agriculture, increasing social and environmental effects, and rising water conflicts between users.

Progress on drought preparedness and policy development has been slow for a number of reasons. It certainly relates to the slow-onset characteristics of drought and the lack of a universal definition. These characteristics (defined in more detail in Chapter 1) make early warning, impact assessment, and response difficult for scientists, natural resource managers, and policy makers. The lack of a universal definition often leads to confusion and inaction on the part of decision makers because scientists may disagree on the existence of drought conditions and severity. Severity is also difficult to characterize because it is best evaluated on the basis of multiple indicators and indices rather than a single variable. The impacts of drought are also largely nonstructural and spatially extensive, making it difficult to assess the effects of drought and respond in a timely and effective manner. Drought and its impacts are not as visual as other natural hazards, making it difficult for the media to communicate the significance of the event and its impacts to the public. Public sentiment to respond is often lacking in comparison to other natural hazards that result in loss of life and property.

Another constraint to drought preparedness has been the dearth of methodologies available to planners to guide them through the planning process. Drought differs in its characteristics between climate regimes, and impacts are locally defined by unique economic, social, and environmental characteristics. A methodology developed by Wilhite (1991) and revised to incorporate greater emphasis on risk management (Wilhite et al., 2000) has provided a set of guidelines or a checklist of the key elements of a drought plan and a process through which they can be adapted to any level of government (i.e., local, state or provincial, or national) or geographical setting as part of a natural disaster or sustainable development plan, an integrated water resources plan, or a standalone drought mitigation plan. We describe this process here, with the goal of providing a template that government or organizations can follow to reduce societal vulnerability to drought.

II. PLANNING FOR DROUGHT: THE PROCESS

Drought is a natural hazard that differs from other hazards in that it has a slow onset, evolves over months or even years, affects a large spatial region, and causes little structural damage. Its onset and end are often difficult to determine, as is its severity. Like other hazards, the impacts of drought span economic, environmental, and social sectors and can be reduced through mitigation and preparedness. Because droughts are a normal part of climate variability for virtually all regions, it is important to develop drought preparedness plans to deal with these extended periods of water shortage in a timely, systematic manner as they evolve. To be effective, these plans must evaluate a region's exposure and vulnerability to the hazard and incorporate these elements in a way that evolves with societal changes.

The 10-step drought planning process developed by Wilhite (1991) was based largely on interactions with many states in the United States, incorporating their experiences and lessons learned. This planning process has gone through several iterations in recent years in order to tailor it to specific countries or subsets of countries (Wilhite et al., 2000). It has also been the basis for discussions at a series of regional training workshops and seminars on drought management and preparedness held throughout the world over the past decade. With the increased interest in drought mitigation planning in recent years, this planning process has evolved to incorporate more emphasis on risk assessment and mitigation tools.

The 10-step drought planning process is illustrated in Figure 2. In brief, Steps 1–4 focus on making sure the right people are brought together, have a clear understanding of the process, know what the drought plan must accomplish, and are supplied with adequate data to make fair and equitable decisions when formulating and writing the actual drought plan. Step 5 describes the process of developing an organizational structure for completion of the tasks necessary to prepare the plan. The plan should be viewed as a process, rather than a discrete event that produces a static document. A risk assessment is undertaken in conjunction with this step in order to construct a vulnerability profile for key economic sectors, population groups, regions, and communities. Steps 6 and 7 detail the need for ongoing research and coordination between scientists and policy makers. Steps 8 and 9 stress the importance of promoting and testing the plan before drought occurs. Finally, Step 10 emphasizes revising the plan to keep it current and evaluating its effectiveness in the post-drought period. Although the steps are sequential, many of these tasks are addressed simultaneously under the leadership of a drought task force and its complement of committees

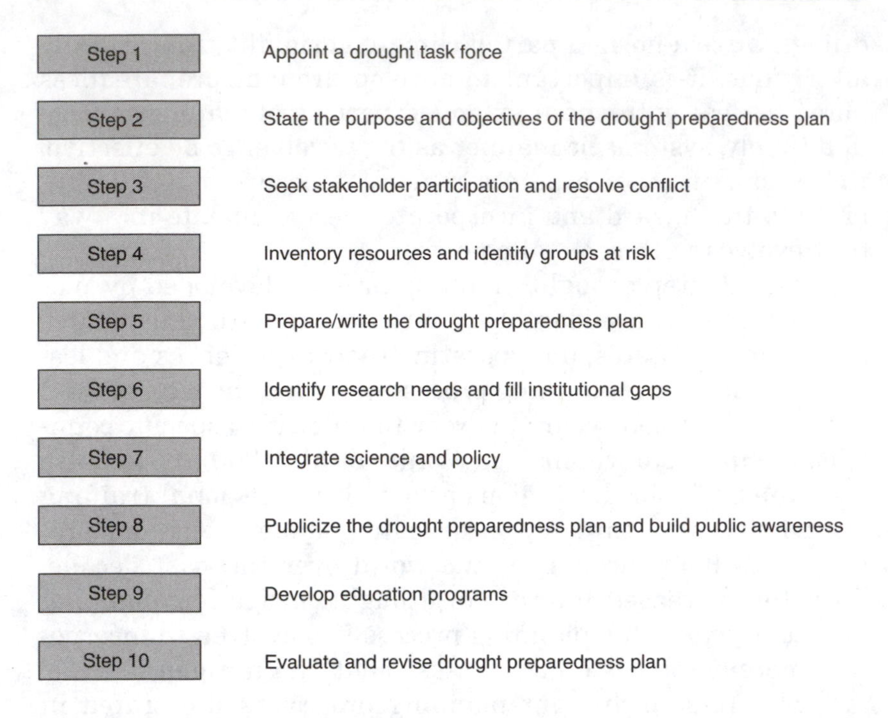

Figure 2 Ten-step planning process. (*Source*: National Drought Mitigation Center, University of Nebraska, Lincoln, Nebraska, USA.)

and working groups. These steps, and the tasks included in each, provide a "checklist" that should be considered and may be completed as part of the planning process.

III. STEP 1: APPOINT A DROUGHT TASK FORCE

A key political leader initiates the drought planning process through appointment of a drought task force. Depending on the level of government developing the plan, this could be the president or prime minister, a provincial or state governor, or a mayor. The task force has two purposes. First, it supervises and coordinates development of the plan. Second, after the plan is developed and during times of drought when the plan is activated, the task force coordinates actions, implements mitigation and response programs, and makes policy recommendations to the appropriate political leader.

The task force should reflect the multidisciplinary nature of drought and its impacts, and it should include appropriate representatives of government agencies (provincial, federal) and universities where appropriate expertise is available. If applicable, the governor's office should have a representative on the task force. Environmental and public interest groups and others from the private sector can be included (see Step 3), as appropriate. These groups would be involved to a considerable extent in the activities of the working groups associated with the Risk Assessment Committee discussed in Step 5. The actual makeup of this task force would vary considerably, depending on the principal economic and other sectors affected, the political infrastructure, and other factors. The task force should include a public information official that is familiar with local media's needs and preferences and a public participation practitioner who can help establish a process that includes and accommodates both well-funded and disadvantaged stakeholder or interest groups.

IV. STEP 2: STATE THE PURPOSE AND OBJECTIVES OF THE DROUGHT PLAN

As its first official action, the drought task force should state the general purpose for the drought plan. Government officials should consider many questions as they define the purpose of the plan, such as the

- Purpose and role of government in drought mitigation and response efforts
- Scope of the plan
- Most drought-prone areas of the state or nation
- Historical impacts of drought
- Historical response to drought
- Most vulnerable economic and social sectors
- Role of the plan in resolving conflict between water users and other vulnerable groups during periods of shortage

- Current trends (e.g., land and water use, population growth) that may increase or decrease vulnerability and conflicts in the future
- Resources (human and economic) the government is willing to commit to the planning process
- Legal and social implications of the plan
- Principal environmental concerns caused by drought

A generic statement of purpose for a plan is to reduce the impacts of drought by identifying principal activities, groups, or regions most at risk and developing mitigation actions and programs that alter these vulnerabilities. The plan is directed at providing government with an effective and systematic means of assessing drought conditions, developing mitigation actions and programs to reduce risk in advance of drought, and developing response options that minimize economic stress, environmental losses, and social hardships during drought.

The task force should then identify the specific objectives that support the purpose of the plan. Drought plan objectives will vary within and between countries and should reflect the unique physical, environmental, socioeconomic, and political characteristics of the region in question. For a provincial, state, or regional plan, objectives that should be considered include the following:

- Collect and analyze drought-related information in a timely and systematic manner.
- Establish criteria for declaring drought emergencies and triggering various mitigation and response activities.
- Provide an organizational structure and delivery system that ensures information flow between and within levels of government.
- Define the duties and responsibilities of all agencies with respect to drought.
- Maintain a current inventory of government programs used in assessing and responding to drought emergencies.

- Identify drought-prone areas of the state or region and vulnerable economic sectors, individuals, or environments.
- Identify mitigation actions that can be taken to address vulnerabilities and reduce drought impacts.
- Provide a mechanism to ensure timely and accurate assessment of drought's impacts on agriculture, industry, municipalities, wildlife, tourism and recreation, health, and other areas.
- Keep the public informed of current conditions and response actions by providing accurate, timely information to media in print and electronic form (e.g., via TV, radio, and the World Wide Web).
- Establish and pursue a strategy to remove obstacles to the equitable allocation of water during shortages and establish requirements or provide incentives to encourage water conservation.
- Establish a set of procedures to continually evaluate and exercise the plan and periodically revise the plan so it will stay responsive to the needs of the area.

V. STEP 3: SEEK STAKEHOLDER PARTICIPATION AND RESOLVE CONFLICT

Social, economic, and environmental values often clash as competition for scarce water resources intensifies. Therefore, it is essential for task force members to identify all citizen groups (stakeholders) that have a stake in drought planning and understand their interests. These groups must be involved early and continuously for fair representation and effective drought management and planning. Discussing concerns early in the process gives participants a chance to develop an understanding of one another's various viewpoints and generate collaborative solutions. Although the level of involvement of these groups will vary notably from location to location, the power of public interest groups in policy making is considerable. In fact, these groups are likely to impede progress in the development of plans if they are not included in the process. The task force should also protect the interests of stakeholders who may lack the financial resources to serve as their own advocates. One way to facilitate public participation is to establish a citizen's advisory council as a permanent feature of the drought plan, to help the task force keep information flowing and resolve conflicts between stakeholders.

State or provincial governments need to consider if district or regional advisory councils should be established. These councils could bring neighbors together to discuss their water use issues and problems and seek collaborative solutions. At the provincial level, a representative of each district council should be included in the membership of the provincial citizens' advisory council to represent the interests and values of their constituencies. The provincial citizens' advisory council can then make recommendations and express concerns to the task force as well as respond to requests for situation reports and updates.

VI. STEP 4: INVENTORY RESOURCES AND IDENTIFY GROUPS AT RISK

An inventory of natural, biological, and human resources, including the identification of constraints that may impede the planning process, may need to be initiated by the task force. In many cases, various provincial and federal agencies already possess considerable information about natural and biological resources. It is important to determine the vulnerability of these resources to periods of water shortage that result from drought. The most obvious *natural resource* of importance is water: its location, accessibility, and quality. *Biological resources* refer to the quantity and quality of grasslands or rangelands, forests, wildlife, and so forth. *Human resources* include the labor needed to develop water resources, lay pipeline, haul water and livestock feed, process citizen complaints, provide technical assistance, and direct citizens to available services.

It is also imperative to identify constraints to the planning process and to the activation of the various elements of the plan as drought conditions develop. These constraints may be physical, financial, legal, or political. The costs associated with plan development must be weighed against the losses that will likely result if no plan is in place. The purpose of a drought plan is to reduce risk and, therefore, economic, social, and environmental impacts. Legal constraints can include water rights, existing public trust laws, requirements for public water suppliers, liability issues, and so forth.

In drought planning, making the transition from crisis to risk management is difficult because, historically, little has been done to understand and address the risks associated with drought. To solve this problem, areas of high risk should be identified, as should actions that can be taken to reduce those risks before a drought occurs. Risk is defined by both the exposure of a location to the drought hazard and the vulnerability of that location to periods of drought-induced water shortages (Blaikie et al., 1994). Drought is a natural event; it is important to define the exposure (i.e., frequency of drought of various intensities and durations) of various parts of the state or region to the drought hazard. Some areas are likely to be more at risk than others. Vulnerability, on the other hand, is affected by social factors such as population growth and migration trends, urbanization, changes in land use, government policies, water use trends, diversity of economic base, cultural composition, and so forth. The drought task force should address these issues early in the planning process so they can provide more direction to the committees and working groups that will be developed under Step 5 of the planning process.

VII. STEP 5: ESTABLISH AND WRITE DROUGHT PLAN

This step describes the process of establishing relevant committees to develop and write the drought plan. The drought plan should have three primary components: (1) monitoring, early warning, and prediction; (2) risk and impact assessment; and (3) mitigation and response. We recommended that a committee be established to focus on the first two of these needs; the drought task force can in most instances carry out the mitigation and response function.

The suggested organizational structure for the plan is illustrated in Figure 3. The committees will have their own tasks and goals, but well-established communication and information flow between committees and the task force is necessary to ensure effective planning.

A. Monitoring, Early Warning, and Prediction Committee

A reliable assessment of water availability and its outlook for the near and long term is valuable information in both dry and wet periods. During drought, the value of this information increases markedly. The monitoring committee should include representatives from agencies with responsibilities for monitoring climate and water supply. Data and information on each of the applicable indicators (e.g., precipitation, temperature, evapotranspiration, seasonal climate forecasts, soil

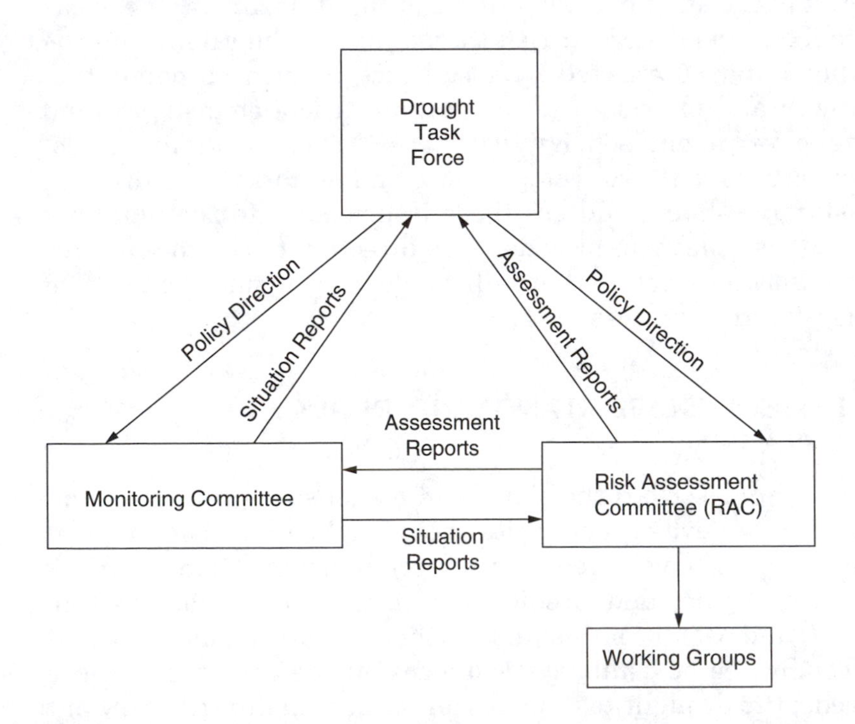

Figure 3 Drought task force organizational structure. (*Source*: National Drought Mitigation Center, University of Nebraska, Lincoln, Nebraska, USA.)

moisture, streamflow, groundwater levels, reservoir and lake levels, and snowpack) ought to be considered in the committee's evaluation of the water situation and outlook. The agencies responsible for collecting, analyzing, and disseminating data and information will vary considerably from country to country and province to province.

The monitoring committee should meet regularly, especially in advance of the peak demand season. Following each meeting, reports should be prepared and disseminated to the drought task force, relevant government agencies, and the media. The chairperson of the monitoring committee should be a permanent member of the drought task force. If conditions warrant, the task force should brief the governor or appropriate government official about the contents of the report, including any recommendations for specific actions. The public must receive a balanced interpretation of changing conditions. The monitoring committee should work closely with public information specialists to keep the public well-informed.

The primary objectives of the monitoring committee are to

1. Adopt a workable definition of drought that could be used to phase in and phase out levels of local state or provincial, and federal actions in response to drought. The group may need to adopt more than one definition of drought in identifying impacts in various economic, social, and environmental sectors because no single definition of drought applies in all cases. Several indices are available (Hayes, 1998), including the Standardized Precipitation Index (McKee et al., 1993, 1995), which is gaining widespread acceptance (Guttman, 1998; Hayes et al., 1999; also see *http://drought. unl.edu/ whatis/Indices.pdf*.
   The trend is to rely on multiple drought indices to trigger mitigation and response actions, which are calibrated to various intensities of drought. The current thought is that no single index of drought is adequate to measure the complex interrelationships between the various components of the hydrological cycle and impacts.

It is helpful to establish a sequence of descriptive terms for water supply alert levels, such as "advisory," "alert," "emergency," and "rationing" (as opposed to more generic terms such as "phase 1" and "phase 2," or sensational terms such as "disaster"). Review the terminology used by other entities (i.e., local utilities, provinces, river basin authorities) and choose terms that are consistent so as not to confuse the public with different terms in areas where there may be authorities with overlapping regional responsibilities. These alert levels should be defined in discussions with both the risk assessment committee and the task force.

In considering emergency measures such as rationing, remember that the impacts of drought may vary significantly from one area to the next, depending on the sources and uses of water and the degree of planning previously implemented. For example, some cities may have recently expanded their water supply capacity while other adjacent communities may have an inadequate water supply capacity during periods of drought. Imposing general emergency measures on people or communities without regard for their existing vulnerability may result in political repercussions and loss of credibility.

A related consideration is that some municipal water systems may be out of date or in poor operating condition, so that even moderate drought strains a community's ability to supply customers with water. Identifying inadequate (i.e., vulnerable) water supply systems and upgrading those systems should be part of a long-term drought mitigation program.

2. Establish drought management areas; that is, subdivide the province or region into more conveniently sized districts by political boundaries, shared hydrological characteristics, climatological characteristics, or other means such as drought probability or risk. These subdivisions may be useful in drought management because they may allow drought stages and mitigation and response options to be regionalized.

3. Develop a drought monitoring system. The quality of meteorological and hydrological networks is highly variable from country to country and region to region within countries. Responsibility for collecting, analyzing, and disseminating data is divided between many government authorities. The monitoring committee's challenge is to coordinate and integrate the analysis so decision makers and the public receive early warning of emerging drought conditions.
   Considerable experience has developed in recent years with automated weather data networks that provide rapid access to climate data. These networks can be invaluable in monitoring emerging and ongoing drought conditions. Investigate the experiences of regions with comprehensive automated meteorological and hydrological networks and apply their lessons learned, where appropriate.
4. Inventory data quantity and quality from current observation networks. Many networks monitor key elements of the hydrologic system. Most of these networks are operated by federal or provincial agencies, but other networks also exist and may provide critical information for a portion of a province or region. Meteorological data are important but represent only one part of a comprehensive monitoring system. These other physical indicators (soil moisture, streamflow, reservoir and groundwater levels) must be monitored to reflect impacts of drought on agriculture, households, industry, energy production, transportation, recreation and tourism, and other water users.
5. Determine the data needs of primary users. Developing new or modifying existing data collection systems is most effective when the people who will be using the data are consulted early and often. Soliciting input on expected new products or obtaining feedback on existing products is critical to ensuring that products meet the needs of primary users and, therefore, will be used in decision making. Training on how

to use or apply products in routine decision making is also essential.

6. Develop or modify current data and information delivery systems. People need to be warned of drought as soon as it is detected, but often they are not. Information needs to reach people in time for them to use it in making decisions. In establishing information channels, the monitoring committee needs to consider when people need what kinds of information. These decision points can determine whether the information provided is used or ignored.

B. Risk Assessment Committee

Risk is the product of exposure to the drought hazard (i.e., probability of occurrence) and societal vulnerability, represented by a combination of economic, environmental, and social factors. Therefore, to reduce vulnerability to drought, one must identify the most significant impacts and assess their underlying causes. Drought impacts cut across many sectors and across normal divisions of government authority. These impacts have been classified by Wilhite and Vanyarkho (2000) and are available on the website of the National Drought Mitigation Center (NDMC) (*http://drought.unl.edu*).

The membership of the risk assessment committee should represent economic sectors, social groups, and ecosystems most at risk from drought. The committee's chairperson should be a member of the drought task force. Experience has demonstrated that the most effective approach to follow in determining vulnerability to and impacts of drought is to create a series of working groups under the aegis of the risk assessment committee. The responsibility of the committee and working groups is to assess sectors, population groups, communities, and ecosystems most at risk and identify appropriate and reasonable mitigation measures to address these risks. Working groups would be composed of technical specialists representing those areas referred to above. The chair of each working group, as a member of the risk assessment committee, would report directly to the committee. Following this model, the responsibility of the risk assessment committee is to direct the activities of each of the working groups and make recommendations to the drought task force on mitigation actions.

The number of working groups will vary considerably between countries or provinces, reflecting the principal impact sectors. The more complex the economy and society, the larger the number of working groups is necessary to reflect these sectors. Working groups may focus on some combination of the following sectors: agriculture, recreation and tourism, industry, commerce, drinking water supplies, energy, environment, wildfire protection, and health.

In drought management, making the transition from crisis to risk management is difficult because little has been done to understand and address the risks associated with drought. A methodology has been developed by the NDMC to help guide drought planners through the risk assessment process. This methodology focuses on identifying and ranking the priority of relevant drought impacts; examining the underlying environmental, economic, and social causes of these impacts; and then choosing actions that will address these underlying causes. What makes this methodology different and more helpful than previous methodologies is that it addresses the causes behind drought impacts. Previously, responses to drought have been reactions to impacts. Understanding why specific impacts occur provides the opportunity to lessen impacts in the future by addressing these vulnerabilities through the identification and adoption of specific mitigation actions. This methodology is described below, divided into six specific tasks. Once the risk assessment committee identifies the working groups, each of these groups would follow this methodology.

1. Task 1: Assemble the Team

It is essential to bring together the right people and supply them with adequate data to make fair, efficient, and informed decisions pertaining to drought risk. Members of this group should be technically trained in the specific topical areas covered by the working groups. When dealing with the issues of appropriateness, urgency, equity, and cultural awareness in drought risk analysis, include public input and consideration. Public participation could be warranted at every step, but time and money may limit involvement to key stages in the risk analysis and planning process (public review vs. public participation). The amount of public involvement is at the discretion of the drought task force and other members of the planning team. The advantage of publicly discussing questions and options is that the procedures used in making any decision will be better understood, and it will also demonstrate a commitment to participatory management. At a minimum, decisions and reasoning should be openly documented to build public trust and understanding.

The choice of specific actions to deal with the underlying causes of the drought impacts will depend on the economic resources available and related social values. Typical concerns are associated with cost and technical feasibility, effectiveness, equity, and cultural perspectives. This process has the potential to lead to the identification of effective and appropriate drought risk reduction activities that will reduce long-term drought impacts, rather than ad hoc responses or untested mitigation actions that may not effectively reduce the impact of future droughts.

2. Task 2: Drought Impact Assessment

Impact assessment examines the consequences of a given event or change. For example, drought is typically associated with a number of outcomes. Drought impact assessments begin by identifying direct consequences of the drought, such as reduced crop yields, livestock losses, and reservoir depletion. These direct outcomes can then be traced to secondary consequences (often social effects), such as the forced sale of household assets or land, dislocation, or physical and emotional stress. This initial assessment identifies drought impacts but does not identify the underlying reasons for these impacts.

Drought impacts can be classified as economic, environmental, or social, although many impacts may span more than one sector. Table1 provides a detailed checklist of impacts that could affect a region or location. Recent drought impacts, especially if they are associated with severe to extreme drought, should be weighted more heavily than the impacts of historical drought, in most cases. Recent events more accurately reflect current vulnerabilities, the purpose of this exercise. Attention should also be given to specific impacts that are expected to emerge in the future.

To perform an assessment using the checklist in Table 1, check the box in front of each category that has been affected by drought in your study area. Classify the types of impacts according to the severity of drought, noting that in the future, droughts of lesser magnitude may produce more serious impacts if vulnerability is increasing. Hopefully, interventions taken now will reduce these vulnerabilities in the future. Define the "drought of record" for each region. Droughts differ from one another according to intensity, duration, and spatial extent. Thus, there may be several droughts of record, depending on the criteria emphasized (i.e., most severe drought of a season or year vs. most severe multi-year drought). These analyses would yield a range of impacts related to the severity of drought. In addition, highlighting past, current, and potential impacts may reveal trends that will also be useful for planning purposes. These impacts highlight sectors, populations, or activities that are vulnerable to drought and, when evaluated with the probability of drought occurrence, identify varying levels of drought risk.

3. Task 3: Ranking Impacts

After each working group has completed the checklist in Table 1, the unchecked impacts should be omitted. This new list will contain the relevant drought impacts for your location or activity. From this list, prioritize impacts according to what work group members consider to be the most important. To be effective and equitable, the ranking should consider concerns such as cost, areal extent, trends over time, public

Table 1 Checklist of Historical, Current, and Potential Drought Impacts

To perform an assessment using this checklist, check the box in front of each category that has been affected by drought in your study area. Your selections can be based on common or extreme droughts, or a combination of the two. For example, if your drought planning was going to be based on the "drought of record," you would need to complete a historical review to identify the drought of record for your area and assess the impacts of that drought. You would then record the impacts on this checklist by marking the appropriate boxes under the "historical" column. Next, with the knowledge you have about your local area, if another drought of record were to occur tomorrow, consider what the local impacts may be and record them on the checklist under the "current" column. Finally, consider possible impacts of the same drought for your area in 5 or 10 years and record these in the "potential" column.

If enough time, money, and personnel are available, it may be beneficial to conduct impact studies based on common droughts, extreme droughts, and the drought of record for your region. These analyses would yield a range of impacts related to the severity of the drought, which is necessary for conducting Step 3 of the guide and could be useful for planning purposes.

H = historical drought
C = current drought
P = potential drought

H	C	P	Economic Impacts
			Loss from crop production
☐	☐	☐	Annual and perennial crop losses
☐	☐	☐	Damage to crop quality
☐	☐	☐	Reduced productivity of cropland (wind erosion, etc.)
☐	☐	☐	Insect infestation
☐	☐	☐	Plant disease
☐	☐	☐	Wildlife damage to crops
			Loss from dairy and livestock production
☐	☐	☐	Reduced productivity of rangeland
☐	☐	☐	Forced reduction of foundation stock
☐	☐	☐	Closure/limitation of public lands to grazing
☐	☐	☐	High cost/unavailability of water for livestock
☐	☐	☐	High cost/unavailability of feed for livestock
☐	☐	☐	High livestock mortality rates
☐	☐	☐	Disruption of reproduction cycles (breeding delays or unfilled pregnancies)

Table 1 Checklist of Historical, Current, and Potential Drought Impacts (continued)

☐	☐	☐	Decreased stock weights
☐	☐	☐	Increased predation
☐	☐	☐	Range fires
			Loss from timber production
☐	☐	☐	Wildland fires
☐	☐	☐	Tree disease
☐	☐	☐	Impaired productivity of forest land
			Loss from fishery production
☐	☐	☐	Damage to fish habitat
☐	☐	☐	Loss of young fish due to decreased flows
☐	☐	☐	Income loss for farmers and others directly affected
☐	☐	☐	Loss of farmers through bankruptcy
☐	☐	☐	Unemployment from drought-related production declines
☐	☐	☐	Loss to recreational and tourism industry
☐	☐	☐	Loss to manufacturers and sellers of recreational equipment
☐	☐	☐	Increased energy demand and reduced supply because of drought-related power curtailments
☐	☐	☐	Costs to energy industry and consumers associated with substituting more expensive fuels (oil) for hydroelectric power
☐	☐	☐	Loss to industries directly dependent on agricultural production (e.g., machinery and fertilizer manufacturers, food processors, etc.)
			Decline in food production/disrupted food supply
☐	☐	☐	Increase in food prices
☐	☐	☐	Increased importation of food (higher costs)
☐	☐	☐	Disruption of water supplies
			Revenue to water supply firms
☐	☐	☐	Revenue shortfalls
☐	☐	☐	Windfall profits
☐	☐	☐	Strain on financial institutions (foreclosures, greater credit risks, capital shortfalls, etc.)
☐	☐	☐	Revenue losses to federal, state, and local governments (from reduced tax base)
☐	☐	☐	Loss from impaired navigability of streams, rivers, and canals
☐	☐	☐	Cost of water transport or transfer

Table 1 Checklist of Historical, Current, and Potential Drought Impacts (continued)

H	C	P	
☐	☐	☐	Cost of new or supplemental water resource development
☐	☐	☐	Cost of increased groundwater depletion (mining), land subsidence
☐	☐	☐	Reduction of economic development
☐	☐	☐	Decreased land prices
			Damage to animal species
☐	☐	☐	Reduction and degradation of fish and wildlife habitat
☐	☐	☐	Lack of feed and drinking water
☐	☐	☐	Disease
☐	☐	☐	Increased vulnerability to predation (from species concentration near water)
☐	☐	☐	Migration and concentration (loss of wildlife in some areas and too many in others)
☐	☐	☐	Increased stress to endangered species
H	**C**	**P**	**Environmental Impacts**
☐	☐	☐	Damage to plant species
☐	☐	☐	Increased number and severity of fires
☐	☐	☐	Loss of wetlands
☐	☐	☐	Estuarine impacts (e.g., changes in salinity levels)
☐	☐	☐	Increased groundwater depletion, land subsidence
☐	☐	☐	Loss of biodiversity
☐	☐	☐	Wind and water erosion of soils
☐	☐	☐	Reservoir and lake levels
☐	☐	☐	Reduced flow from springs
☐	☐	☐	Water quality effects (e.g., salt concentration, increased water temperature, pH, dissolved oxygen, turbidity)
☐	☐	☐	Air quality effects (e.g., dust, pollutants)
☐	☐	☐	Visual and landscape quality (e.g., dust, vegetative cover, etc.)
H	**C**	**P**	**Social Impacts**
☐	☐	☐	Mental and physical stress (e.g., anxiety, depression, loss of security, domestic violence)
☐	☐	☐	Health-related low-flow problems (e.g., cross-connection contamination, diminished sewage flows, increased pollutant concentrations, reduced firefighting capability, etc.)

TABLE 1 Checklist of Historical, Current, and Potential Drought Impacts (continued)

☐	☐	☐	Reductions in nutrition (e.g., high-cost-food limitations, stress-related dietary deficiencies)
☐	☐	☐	Loss of human life (e.g., from heat stress, suicides)
☐	☐	☐	Public safety from forest and range fires
☐	☐	☐	Increased respiratory ailments
☐	☐	☐	Increased disease caused by wildlife concentrations
			Increased conflicts
☐	☐	☐	Water user conflicts
☐	☐	☐	Political conflicts
☐	☐	☐	Management conflicts
☐	☐	☐	Other social conflicts (e.g., scientific, media based)
☐	☐	☐	Disruption of cultural belief systems (e.g., religious and scientific views of natural hazards)
☐	☐	☐	Reevaluation of social values (e.g., priorities, needs, rights)
☐	☐	☐	Reduction or modification of recreational activities
☐	☐	☐	Public dissatisfaction with government regarding drought response
☐	☐	☐	Inequity in the distribution of drought relief
			Inequity in drought impacts based on:
☐	☐	☐	Socioeconomic group
☐	☐	☐	Ethnicity
☐	☐	☐	Age
☐	☐	☐	Gender
☐	☐	☐	Seniority
☐	☐	☐	Loss of cultural sites
☐	☐	☐	Loss of aesthetic values
☐	☐	☐	Recognition of institutional restraints on water use
			Reduced quality of life, changes in lifestyle
☐	☐	☐	In rural areas
☐	☐	☐	In specific urban areas
☐	☐	☐	Increased poverty in general
☐	☐	☐	Increased data/information needs, coordination of dissemination activities
☐	☐	☐	Population migrations (e.g., rural to urban areas, migrants into the United States)

opinion, fairness, and the ability of the affected area to recover. Be aware that social and environmental impacts are often difficult, if not impossible, to quantify. Each work group should complete a preliminary ranking of impacts. The drought task force and other work groups can participate in a plenary discussion of these rankings following the initial ranking iterations. We recommend constructing a matrix (see an example in Table 2) to help prioritize impacts. From this list of prioritized impacts, each working group should decide which impacts should be addressed and which can be deferred.

4. Task 4: Vulnerability Assessment

Vulnerability assessment provides a framework for identifying the social, economic, and environmental causes of drought impacts. It bridges the gap between impact assessment and policy formulation by directing policy attention to underlying causes of vulnerability rather than the result, the negative impacts, which follow triggering events such as drought (Ribot et al., 1996). For example, the direct impact of precipitation deficiencies may be a reduction of crop yields. The underlying cause of this vulnerability, however, may be that the farmers did not use drought-resistant seeds, because they did not believe in their usefulness, the costs were too high, or there was some commitment to cultural beliefs. Another example could be farm foreclosure. The underlying causes of this vulnerability might include small

Table 2 Drought Impact Decision Matrix

Impact	Cost	Equally Distributed?	Growing?	Public Priority?	Equitable Recovery?	Impact Rank

farm size because of historical land appropriation policies, lack of credit for diversification options, farming on marginal lands, limited knowledge of possible farming options, a lack of local industry for off-farm supplemental income, or government policies. Therefore, for each of the identified impacts from Table 1, begin asking why these impacts have occurred or might occur. It is important to realize that a combination of factors might produce a given impact. It might be beneficial to display these causal relationships in some form of a tree diagram (see examples in Figures 4 and 5). Figure 4 demonstrates a typical agricultural example and Figure 5 a potential urban scenario. Depending on the level of analysis, this process can quickly become somewhat complicated, which is why working groups must be composed of

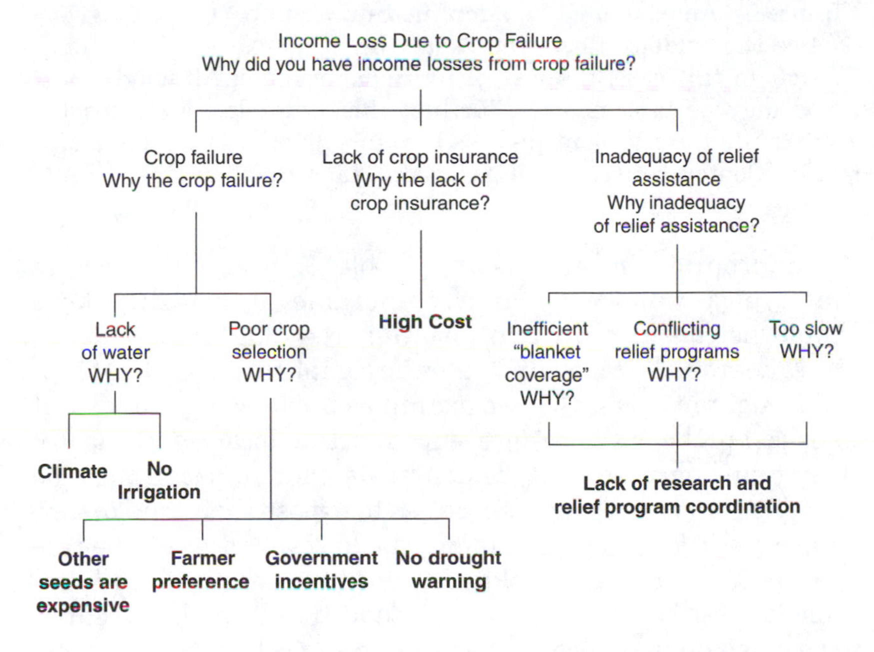

Figure 4 An example of a simplified agricultural impact tree diagram. (Notice the boldface items represent the basal causes of the listed impact. Although these items may be broken down further, this example illustrates the vulnerability assessment process.) (*Source*: National Drought Mitigation Center, University of Nebraska, Lincoln, Nebraska, USA.)

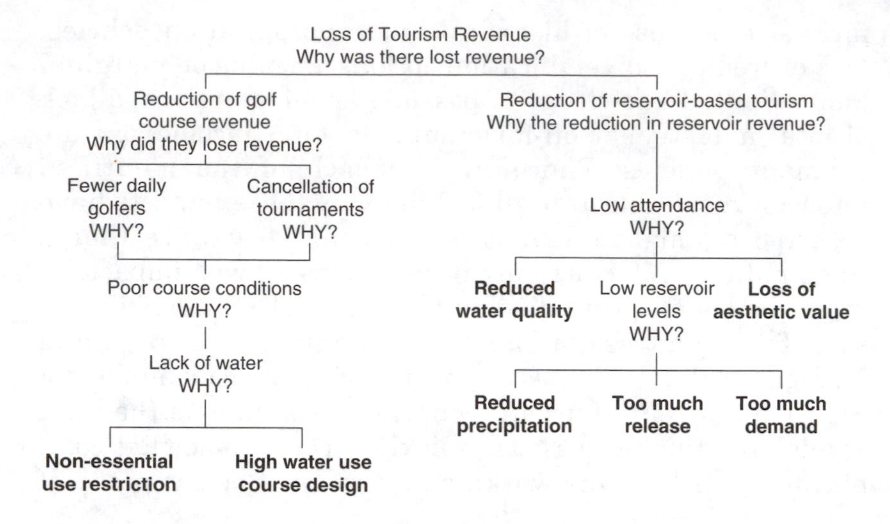

Figure 5 An example of a simplified urban impact tree diagram. (Notice the boldface items represent the basal causes of the listed impact [in this case, the loss of tourism revenue]. Although these items may be broken down further, this example illustrates the vulnerability assessment process.) (*Source*: National Drought Mitigation Center, University of Nebraska, Lincoln, Nebraska, USA.)

the appropriate mix of people. Table 3 lists many factors that typically make an area vulnerable to drought; these should be considered when forming tree diagrams.

The tree diagrams illustrate the complexity of understanding drought impacts. The two examples provided are not meant to be comprehensive or represent an actual location. Basically, their main purpose is to demonstrate that impacts must be examined from several perspectives to expose their true underlying causes. For this assessment, the lowest causes on the tree diagrams, the items in boldface, will be referred to as basal causes. These basal causes are the items that have the potential to be acted on to reduce the associated impact. Of course, some of these impact causes should not or cannot be acted on for a wide variety of reasons (discussed in Step 5).

5. Task 5: Action Identification

Mitigation is defined as actions taken in advance of or in the early stages of drought that reduce the impacts of the event.

TABLE 3 Vulnerability Consideration

	Water Shortage Vulnerability Continuum	
	Higher Vulnerability	Lower Vulnerability
Meteorological drought	Wide precipitation variation	Stable precipitation pattern
	Lack of data/single-source data	Good long-term data/multiple sources of data
	Passive drought "acceptance"	Advance warning
	Longer duration	Shorter duration
	Higher severity shortage	Lower severity shortage
	Sudden change in supply	Gradual changes in supply
Supply-and-demand balance or "institutional drought"	Single water source or low supply reliability	Multiple water sources or high supply reliability
	Low-priority water rights or low contractual rights	Senior water rights or high contractual rights
	Water supply at risk from contamination	Protected water supply
	Imported water supplies	Local supplies and locally controlled
	Subject to other natural disasters	Low likelihood of other natural disasters
Water use patterns	High-growth area/high additional demand	Stable or decreasing water demand
	High percent water use improvements requires earlier demand management response	Low percent water use/efficiency improves "slack" in system = requires more demand management response
	Landscape/ag irrigation usual practices or landscape/ag dependence on precipitation	"Climate-appropriate" plants or nonirrigated agriculture/grazing
Preparedness	Wait until shortage is "declared" (or beyond ...)	Early shortage response

TABLE 3 Vulnerability Consideration (continued)

Water Shortage Vulnerability Continuum	
Higher Vulnerability	Lower Vulnerability
Lack of political will	Leadership
Ignoring situation/abdicating responsibility	Preparedness/actions to protect community/economy/environment
Noninterconnected water supply systems or noncollaborative approach with neighbors	Coordination with others (i.e., neighboring water, disaster response and fire agencies, mutual aid agreements, etc.)
Revenue/rate instability	Rate stabilization fund
"Knee jerk" rationing	Predetermined and equitable allocation methods
Little public awareness	High community involvement (from all social and economic sectors)

Source: D. Braver, personal communication, 1997.

Once the group has set drought impact priorities and exposed the corresponding underlying causes of vulnerability, it can identify actions appropriate for reducing drought risk. The matrix lists the impact as well as the described basal causes of the impact. From this point, the working group should investigate what actions could be taken to address each of these basal causes. The following sequence of questions may be helpful in identifying potential actions:

- Can the basal cause be mitigated (can it be modified before a drought)? If yes, then how?
- Can the basal cause be responded to (can it be modified during or after a drought)? If so, then how?
- Is there some basal cause, or aspect of the basal cause, that cannot be modified and must be accepted as a drought-related risk for this activity or area?

A list of potential actions to mitigate drought is available at *http://drought.unl.edu/plan/handbook/risk.pdf*. As will be discussed in the next section (Task 6), not all ideas are appropriate in all cases. Many of the ideas are more in the realm of short-term emergency response, or crisis management, rather than long-term mitigation, or risk management. Emergency response is an important component of drought planning, but it should be only one part of a more comprehensive mitigation strategy.

6. Task 6: Developing the "To Do" List

After the group identifies the impacts, causes, and relevant potential actions, the next step is to determine the sequence of actions to take as part of the risk reduction planning exercise. This selection should be based on such concerns as feasibility, effectiveness, cost, and equity. Additionally, it will be important to review the impact tree diagrams when considering which groups of actions need to be considered together. For example, if you wanted to reduce crop losses by promoting the use of a different type of seed, it probably would not be effective to educate farmers on the benefits of the new variety if it is too expensive or there are government incentives for planting other crops.

In choosing the appropriate actions, you may want to ask some of the following questions:

- What are the cost/benefit ratios for the actions identified?
- Which actions do the general public consider feasible and appropriate?
- Which actions are sensitive to the local environment (i.e., sustainable practices)?
- Do the actions address the right combination of causes to adequately reduce the relevant impact?
- Do the actions address short- and long-term solutions?
- Which actions would fairly represent the needs of affected individuals and groups?

This process has the potential to lead to the identification of effective and appropriate drought risk reduction activities that may reduce future drought impacts.

7. Completion of Risk Analysis

Following Task 6, the risk analysis is finished. Remember, this is a planning process, so it will be necessary to periodically reevaluate drought risk and the various mitigation actions identified. Step 10 in the planning process is associated with evaluating, testing, and revising the drought plan. After a severe drought episode would be an appropriate time to revisit mitigation actions in association with an analysis of lessons learned.

C. Mitigation and Response Committee

Mitigation and response actions may be the responsibility of the drought task force or be assigned to a separate committee. It is recommended that the task force, working in cooperation with the monitoring and risk assessment committees, has the knowledge and experience to understand drought mitigation techniques, risk analysis (economic, environmental, and social aspects), and drought-related decision-making processes at all levels of government. The task force, as originally defined, is composed of senior policy makers from various government agencies and, possibly, key stakeholder groups. Therefore, it is in an excellent position to recommend or implement mitigation actions, request assistance through various federal programs, or make policy recommendations to a legislative body or political leader.

Mitigation and response actions should be determined for each of the principal impact sectors identified by the risk assessment committee. Wilhite (1997) assessed drought mitigation technologies implemented by U.S. states in response to drought conditions during the late 1980s and early 1990s. The transferability of these technologies to specific settings or locations needs to be evaluated further. These drought technologies are available on the NDMC's website (*http://drought.unl.edu/mitigate/mitigate.htm*).

The State of Georgia recently developed a drought management plan and identified a broad range of pre-drought mitigation strategies that could be used to lessen the state's vulnerability to future drought events. These strategies are divided by sector into municipal and industrial, agriculture, and water quality. Selected examples of these actions are provided in Table 4. These examples illustrate the types of actions identified by states that have recently completed the drought mitigation planning process.

Tribal governments in the United States, many of which are located in extremely drought-prone areas, are also pursuing development of drought mitigation plans. For example, the Hopi Nation followed the 10-step guidelines and the NDMC's risk assessment methodology (Knutson et al., 1998) in its plan development process. The plan is pending approval through the U.S. Bureau of Reclamation, which provided funding for its development. The vulnerability analysis revealed four sectors of concern: range and livestock, agriculture, village water supplies, and environmental health. A unique feature of the Hopi drought plan is the inclusion of current and proposed monitoring systems to evaluate climatic conditions, soil, vegetation, and water resources for farming, ranching, and domestic purposes. The drought plan describes establishing a

Table 4 Summary of Selected Pre-Drought Strategies Included in the Georgia Drought Management Plan

MUNICIPAL AND INDUSTRIAL *State Actions*	AGRICULTURE *Farmer Irrigation Education*	WATER QUALITY, FLORA, AND FAUNA *State Actions*
Formalize the Drought Response Committee as a means of expediting communications among state, local, and federal agencies and nongovernmental entities.	Recommend that farmers attend classes in best management practices (BMP) and conservation irrigation before (i) receiving a permit, (ii) using a new irrigation system, or (iii) irrigating for a coming announced drought season.	Encourage all responsible agencies to promote voluntary water conservation through a wide range of activities.
Establish a drought communications systems between the state and local governments and water systems.	Provide continuing education opportunities for farmers.	Monitor streamflow and precipitation at selected locations on critical streams.
Provide guidance to the local governments and water supply providers on long-term water supply, conservation and drought contingency planning.	Encourage the use of BMPs, conservation irrigation, efficient use of irrigation systems, and the Cooperative Extension Service's water conservation guidelines.	Monitor water quality parameters, such as temperature and dissolved oxygen at selected critical streams.
Review the local governments and water supply providers' conservation and drought contingency plans.	Develop electronic database for communicating with water use permit holders.	Provide the streamflow and water-quality data in real time for use by drought managers and work with drought managers to optimize information delivery and use.

Work with the golf course and turf industry to establish criteria for drought-tolerant golf courses.	Encourage development and distribution of information on water efficient irrigation techniques.	Evaluate the impact of water withdrawals on flow patterns, and the impact of wastewater discharges on water quality during drought.
Encourage water re-use.	***Field/Crop Type Management*** Encourage the use of more drought-resistant crops.	Investigate indicators and develop tools to analyze drought impacts for waterways such as coastal ecosystems, thermal refuges such as the Flint River, and trout streams.
Provide water efficiency education for industry and business.	Encourage the use of innovative cultivation techniques to reduce crop water use.	Improve the agencies' capabilities and resources to monitor land-disturbing activities that might result in erosion and sedimentation violations.
Conduct voluntary water audits for businesses that use water for production of a product or service.	Conduct crop irrigation efficiency studies.	Identify funding mechanisms and develop rescue and reintroduction protocols for threatened and endangered species during extreme events.
Identify vulnerable water-dependent industries, fund research to help determine impacts and improve predictive capabilities.	Provide farmers with normal year, real time irrigation, irrigation scheduling, and crop evaporation/transpiration information.	Develop and execute an effort to identify pollutant load reduction opportunities by wastewater discharge permit holders.

Table 4 Summary of Selected Pre-Drought Strategies Included in the Georgia Drought Management Plan (continued)

MUNICIPAL AND INDUSTRIAL	AGRICULTURE	WATER QUALITY, FLORA, AND FAUNA
State Actions	***Farmer Irrigation Education***	***State Actions***
Develop and implement a statewide water conservation program to encourage local and regional conservation measures.	Irrigation Equipment Management Encourage the installation of water-efficient irrigation technology.	Evaluate the impact of water withdrawals on flow regimes and the impact of wastewater discharges on water quality during drought.
Develop and implement an incentive program to encourage more efficient use of existing water supplies.	Retrofit older irrigation systems with newer and better irrigation technology. Update any system more than 10 years old.	Develop and promote implementation of sustainable lawn care programs based on selected BMPs and/or integrated pest management practices.
Local/Regional Actions Develop and implement a drought management and conservation plan.	Encourage farmers to take advantage of available financial incentives for retrofitting and updating older or less efficient systems.	Encourage protection and restoration of vegetated stream buffers, including incentives for property owners to maintain buffers wider than the minimum required by state law.
Assess and classify drought vulnerability of individual water systems.	Recommend irrigation system efficiency audits every 5 to 7 years.	Provide for protection of recharge areas through measures including land purchase or acquisition of easements.

Define predetermined drought responses, with outdoor watering restrictions being at least as restrictive as the state's minimum requirements.	***Government Programs***	Encourage and explore wildland fire mitigation measures.
	Improve irrigation permit data to create a high degree of confidence in the information on ownership, location, system type, water source, pump capacity, and acres irrigated for all irrigation systems to determine which watersheds and aquifers will be strongly affected by agricultural water use, especially in droughts.	
Establish a drought communications system from local governments and water supply systems to the public.	Improve on the agriculture irrigation water measurement and accounting statewide.	Enhance programs to assist landowners and farmers with outdoor burning.
	Improve communications and cooperation among farmers and relevant state and federal agencies regarding available assistance during drought conditions.	
	Support legislation and efforts to enhance the ability of farmers to secure adequate water supplies during drought conditions.	

Source: Georgia Department of Natural Resources (2003).

network of approximately 60 transects to provide a detailed analysis of range conditions. The transects will be selected to represent major climates, soils, water resources, and land uses present on the Hopi reservation and will help identify trends in vegetation health. These monitoring networks will not only help monitor and quantify the drought impacts, but also be used to assess the effectiveness of any mitigation actions that are implemented.

As was the case with the Georgia plan cited above, the Hopi drought plan developed a list of short- and long-term drought mitigation and response actions for each impact area. For example, to mitigate range and livestock losses, the plan suggests that range management plans be completed for each range unit. To facilitate rotations and proper use of rangelands, the Hopi range management plan also includes fencing and water development projects for the unit range management plans. Water availability in these units will be improved through a combination of rehabilitating surface water impoundments, additional wells at key locations, improved water distribution from the supply point to multiple stock watering troughs, and other conjunctive uses. The Hopi planners hope these mitigation actions will decrease the vulnerability of the range and livestock economic sector.

In addition to identifying mitigation actions that will reduce the tribe's drought risk, the Hopi drought plan is unique in that it identifies the responsible agencies, provides a timeline to complete the actions, and proposes a cost estimate for these actions. For example, a cost of $12 million is estimated to upgrade the water supply systems of 12 tribal villages by improving pumping capacity, storage tank size, and pipe capacity. The tribe plans to seek funding for these actions through a variety of agencies and sources while enhancing water conservation at the same time.

Before the onset of drought, the task force should inventory all forms of assistance available from governmental and nongovernmental authorities during severe drought. The task force should evaluate these programs for their ability to address short-term emergency situations and long-term mitigation programs for their ability to reduce risk to drought. Assistance should be defined in a very broad way to include all forms of technical, mitigation, and relief programs available.

D. Writing the Plan

With input from each of the committees and working groups, the drought task force, with the assistance of professional writing specialists, will draft the drought plan. After completion of a working draft, we recommend holding public meetings or hearings at several locations to explain the purpose, scope, and operational characteristics of the plan. You must also discuss the specific mitigation actions and response measures recommended in the plan. A public information specialist for the drought task force can facilitate planning for the hearings and prepare news stories to announce the meetings and provide an overview of the plan.

As mentioned previously, the plan should not be considered a static document. The plan is dynamic. A copy of the plan should be available through the drought task force website and in hard copy form for distribution.

VIII. STEP 6: IDENTIFY RESEARCH NEEDS AND FILL INSTITUTIONAL GAPS

As research needs and gaps in institutional responsibility become apparent during drought planning, the drought task force should compile a list of those deficiencies and make recommendations to the appropriate person or government body on how to remedy them. You must perform Step 6 concurrently with Steps 4 and 5. For example, the monitoring committee may recommend establishing an automated weather station network or initiating research on the development of a climate or water supply index to help monitor water supplies and trigger specific actions by state government.

IX. STEP 7: INTEGRATE SCIENCE AND POLICY

An essential aspect of the planning process is integrating the science and policy of drought management. The policy maker's understanding of the scientific issues and technical constraints involved in addressing problems associated with drought is often limited. Likewise, scientists generally have a poor understanding of existing policy constraints for responding to the impacts of drought. In many cases, communication and understanding between the science and policy communities must be enhanced if the planning process is to be successful.

Good communication is required between the two groups in order to distinguish what is feasible from what is not achievable for a broad range of science and policy issues. Integration of science and policy during the planning process will also be useful in setting research priorities and synthesizing current understanding. The drought task force should consider various alternatives to bring these groups together and maintain a strong working relationship.

X. STEP 8: PUBLICIZE THE DROUGHT PLAN—BUILD PUBLIC AWARENESS AND CONSENSUS

If you have communicated well with the public throughout the process of establishing a drought plan, there may already be better-than-normal awareness of drought and drought planning by the time you actually write the plan. Themes to emphasize in writing news stories during and after the drought planning process could include:

- How the drought plan is expected to relieve drought impacts in both the short and long term. Stories can focus on the human dimensions of drought, such as how it affects a farm family; on its environmental consequences, such as reduced wildlife habitat; and on its economic effects, such as the costs to a particular industry or to the state or region's overall economy.

- What changes people might be asked to make in response to different degrees of drought, such as restricted lawn watering and car washing or not irrigating certain crops at certain times.

In subsequent years, you may want to do "drought plan refresher" news releases at the beginning of the most drought-sensitive season, letting people know whether there is pressure on water supplies or reason to believe shortfalls will occur later in the season, and reminding them of the plan's existence, history, and any associated success stories. It may be useful to refresh people's memories about circumstances that would lead to water use restrictions.

During drought, the task force should work with public information professionals to keep the public well informed of the status of water supplies, whether conditions are approaching "trigger points" that will lead to requests for voluntary or mandatory use restrictions, and how victims of drought can access assistance. Post all pertinent information on the drought task force's website so that the public can get information directly from the task force without having to rely on mass media.

XI. STEP 9: DEVELOP EDUCATION PROGRAMS

A broad-based education program to raise awareness of short- and long-term water supply issues will help ensure that people know how to respond to drought when it occurs and that drought planning does not lose ground during non-drought years. Try to tailor information to the needs of specific groups (e.g., elementary and secondary education, small business, industry, homeowners, utilities). The drought task force or participating agencies should consider developing presentations and educational materials for events such as a water awareness week, community observations of Earth Day, relevant trade shows, specialized workshops, and other gatherings that focus on natural resource stewardship or management.

XII. STEP 10: EVALUATE AND REVISE DROUGHT PLAN

The final step in the planning process is to create a detailed set of procedures to ensure adequate plan evaluation. Periodic testing, evaluation, and updating of the drought plan are essential to keep the plan responsive to local, state, provincial, or national needs. To maximize the effectiveness of the system, you must include two modes of evaluation: ongoing and post-drought.

A. Ongoing Evaluation

An ongoing or operational evaluation keeps track of how societal changes such as new technology, new research, new laws, and changes in political leadership may affect drought risk and the operational aspects of the drought plan. Drought risk may be evaluated quite frequently whereas the overall drought plan may be evaluated less often. We recommend an evaluation under simulated drought conditions (i.e., drought exercise) before the drought plan is implemented and periodically thereafter. Remember that drought planning is a process, not a discrete event.

B. Post-Drought Evaluation

A post-drought evaluation or audit documents and analyzes the assessment and response actions of government, nongovernmental organizations, and others and provides a mechanism to implement recommendations for improving the system. Without post-drought evaluations, it is difficult to learn from past successes and mistakes, as institutional memory fades.

Post-drought evaluations should include an analysis of the climatic and environmental aspects of the drought; its economic and social consequences; the extent to which predrought planning was useful in mitigating impacts, in facilitating relief or assistance to stricken areas, and in post-recovery; and any other weaknesses or problems caused by or not covered by the plan. Attention must also be directed to situations in which drought-coping mechanisms worked and where societies exhibited resilience; evaluations should not focus only on those situations in which coping mechanisms failed. Evaluations of previous responses to severe drought are also a good planning aid.

To ensure an unbiased appraisal, governments may wish to place the responsibility for evaluating drought and societal response to it in the hands of nongovernmental organizations such as universities or specialized research institutes.

XIII. SUMMARY AND CONCLUSION

For the most part, previous responses to drought in all parts of the world have been reactive, representing the crisis management approach. This approach has been ineffective (i.e., assistance poorly targeted to specific impacts or population groups), poorly coordinated, and untimely; more important, it has done little to reduce the risks associated with drought. In fact, the economic, social, and environmental impacts of drought have increased significantly in recent decades. A similar trend exists for all natural hazards.

This chapter presents a planning process that has been used at all levels of government to guide the development of a drought mitigation plan. The goal of this planning process is to significantly change the way we prepare for and respond to drought by placing greater emphasis on risk management and the adoption of appropriate mitigation actions. The 10 steps included in this process are generic so that governments can choose the steps and components that are most applicable to their situation. The risk assessment methodology is designed to guide governments through the process of evaluating and prioritizing impacts and identifying mitigation actions and tools that can be used to reduce these impacts for future drought episodes. Drought planning must be viewed as an ongoing process, continuously evaluating our changing vulnerabilities and how governments and stakeholders can work in partnership to lessen risk.

REFERENCES

Blaikie, P, T Cannon, I Davis, B Wisner. *At Risk: Natural Hazards, People's Vulnerability, and Disasters*. London: Routledge Publishers, 1994.

Georgia Department of Natural Resources. Georgia Drought Management Plan. Atlanta, GA, 2003. Available online at http://www.state.ga.us/dnr/environ/gaenviron_files/drought_files/drought_mgmtplan_2003.pdf.

Guttman, NB. Comparing the Palmer Drought Index and the Standardized Precipitation Index. *Journal of the American Water Resources Association* 34(1):113–121, 1998.

Hayes, M. *Drought Indices.* Lincoln, NE: National Drought Mitigation Center, University of Nebraska–Lincoln, 1998. Available online at http://drought.unl.edu/whatis/indices.htm.

Hayes, M, M Svoboda, D Wilhite, O Vanyarkho. Monitoring the 1996 drought using the SPI. *Bulletin of the American Meteorological Society* 80:429–438, 1999.

Knutson, C, M Hayes, KT Phillips. How to Reduce Drought Risk. Prepared by the Preparedness and Mitigation Working Group of the Western Drought Coordination Council. Lincoln, NE, 1998. Available online at http://drought.unl.edu/plan/handbook/risk.pdf.

McKee, TB, NJ Doesken, J Kleist. The relationship of drought frequency and duration to time scales. Proceedings of the Eighth Conference on Applied Climatology, Anaheim, CA, January 17–23, American Meteorological Society, Boston, MA, pp. 179–184, 1993.

McKee, TB, NJ Doesken, J Kleist. Drought monitoring with multiple time scales. Proceedings of the Ninth Conference on Applied Climatology, Dallas, TX, January 15–20, American Meteorological Society, Boston, MA, pp. 233–236, 1995.

Ribot, JC, A Najam, G Watson. Climate variation, vulnerability and sustainable development in the semi-arid tropics. In: JC Ribot, AR Magalhães, SS Panagides, eds. *Climate Variability, Climate Change and Social Vulnerability in the Semi-Arid Tropics* (pp. 13–51). New York: Cambridge University Press, 1996.

Wilhite, DA. Drought planning: A process for state government. *Water Resources Bulletin* 27(1):29–38, 1991.

Wilhite, DA. State actions to mitigate drought: Lessons learned. *Journal of the American Water Resources Association* 33(5):961–968, 1997.

Wilhite, DA, O Vanyarkho. Drought: Pervasive impacts of a creeping phenomenon. In: DA Wilhite, ed. *Drought: A Global Assessment* (Volume I, pp. 245–255). London: Routledge Publishers, 2000.

Wilhite, DA, MJ Hayes, C Knutson, KH Smith. Planning for drought: Moving from crisis to risk management. *Journal of the American Water Resources Association* 36:697–710, 2000.

6

National Drought Policy: Lessons Learned from Australia, South Africa, and the United States

DONALD A. WILHITE, LINDA BOTTERILL,
AND KARL MONNIK

CONTENTS

I. INTRODUCTION

Drought is a frequent visitor to Australia, South Africa, and the United States. Each country has struggled to effectively manage drought events, and lessons learned from these attempts have taught these countries that the reactive, crisis management approach is largely ineffective, promoting greater reliance on government and increasing societal vulnerability to subsequent drought episodes. Repeated occurrences of drought in recent decades have placed each nation on a course to develop a national drought policy that promotes improved self-reliance by placing greater emphasis on monitoring and early warning, improving decision support and preparedness planning, and enhancing risk management. Although each nation has differed in its approach, the goal is the same—to reduce societal vulnerability to drought through improved self-reliance while minimizing the need for government intervention.

This chapter describes the process each country has gone through to reach its current level of preparedness and the status of current drought policies. A case study of each country will provide insight into the complexities of the policy development process, the obvious and not-so-obvious pitfalls, and future prospects. The ultimate objective of this chapter is to help other nations achieve a higher level of preparedness and improved drought policy through the transferability of some of the principal lessons learned.

II. DROUGHT POLICY AND PREPAREDNESS: DEFINING A NEW PARADIGM

The implementation of a drought policy can alter a nation's approach to drought management. In the past decade or so, drought policy and preparedness has received increasing attention from governments, international and regional organizations, and nongovernmental organizations. Simply stated, a national drought policy should establish a clear set of principles or operating guidelines to govern the management of drought and its impacts. The policy should be consistent and equitable for all regions, population groups, and economic sectors and consistent with the goals of sustainable development. The overriding principle of drought policy should be an emphasis on risk management through the application of preparedness and mitigation measures. This policy should be directed toward reducing risk by developing better awareness and understanding of the drought hazard and the underlying causes of societal vulnerability. The principles of risk management can be promoted by encouraging the improvement and application of seasonal and shorter term forecasts, developing integrated monitoring and drought early warning systems and associated information delivery systems, developing preparedness plans at various levels of government, adopting mitigation actions and programs, creating a safety net of emergency response programs that ensure timely and targeted relief, and providing an organizational structure that enhances coordination within and between levels of government and with stakeholders.

As vulnerability to drought has increased globally, greater attention has been directed to reducing risks associated with its occurrence through the introduction of planning to improve operational capabilities (i.e., climate and water supply monitoring, building institutional capacity) and mitigation measures aimed at reducing drought impacts. This change in emphasis is long overdue. Mitigating the effects of drought requires the use of all components of the cycle of disaster management (Figure 1), rather than only the crisis management portion of this cycle. Typically, when a natural hazard event and resultant disaster occurs, governments and donors follow with impact assessment, response, recovery, and reconstruction activities to return the region or locality to a pre-disaster state. Historically, little attention has been given to preparedness, mitigation, and prediction or early warning actions (i.e., risk management) that could reduce future impacts and lessen the need for government intervention in the future. Because of this emphasis on crisis management, society has generally moved from one disaster to another with little, if any, reduction in risk. In drought-prone regions, another drought often occurs before the region fully recovers from the last drought.

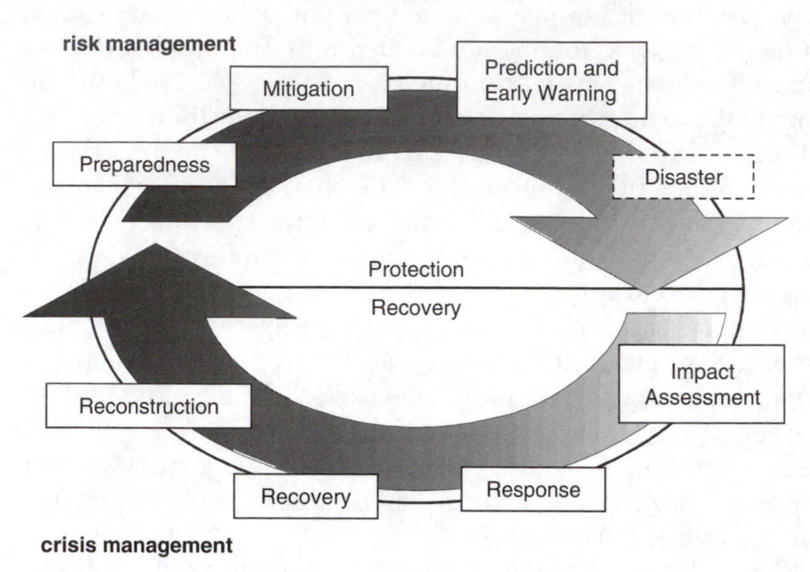

Figure 1 Cycle of disaster management. (Source: National Drought Mitigation Center, University of Nebraska, Lincoln, Nebraska, USA.)

Four key components comprise an effective drought risk reduction strategy: (1) the availability of timely and reliable information on which to base decisions; (2) policies and institutional arrangements that encourage assessment, communication, and application of that information; (3) a suite of appropriate risk management measures for decision makers; and (4) effective and consistent actions by decision makers (O'Meagher et al., 2000). It is critical for governments with drought policy and preparedness experience to share it with other nations that are eager to improve their level of preparedness.

III. NATIONAL DROUGHT POLICY: LESSONS FROM AUSTRALIA

Australia is the driest inhabited continent on earth, and it experiences one of the most variable climates. Unlike other continents, its patterns are determined by nonannual cycles (Flannery, 1994), posing challenges for agricultural practices developed in the relatively more reliable climate of Europe. An early report on the prospects for agriculture in the colony of New South Wales noted the "uncertain climate" and suggested that the future of the colony "will be that of pasture rather than tillage, and the purchase of land will be made with a view to the maintenance of large flocks of fine-woolled sheep; the richer lands, which will generally be found on the banks of the rivers, being devoted to the production of corn, maize and vegetables" (Bigge, 1966, p. 92).

In spite of these early concerns, a successful agricultural industry developed in Australia, becoming the backbone of national prosperity until about the mid-20th century and remaining an important contributor to the country's export earnings. Under Australia's Constitution, agriculture is essentially a state responsibility, with the commonwealth government becoming involved through its fiscal power and by negotiation with the states. This negotiation takes place through the Council of Ministers, first established in 1934 as the Australian Agriculture Council and currently known as the Primary Industries Ministerial Council. The council is supported by a standing committee of senior officials drawn from the commonwealth and state departments responsible for agriculture. The Ministerial Council was the mechanism through which Australia's National Drought Policy was developed and also the forum within which its disputed elements have been fought out.

A. Pre-Drought Policy Period in Australia

Until 1989, drought was considered to be a natural disaster and drought relief was provided in accordance with state disaster relief policy. From the late 1930s, the commonwealth government became progressively more involved in natural disaster relief through a series of ad hoc arrangements with the states and special purpose legislation such as that passed in the mid-1960s to provide drought relief to New South Wales and Queensland.

In 1971, disaster relief arrangements were revised by the commonwealth government and a formula established under which the commonwealth shared the cost of natural disaster relief with the States. This arrangement has continued with a number of minor administrative amendments. In 1989, the commonwealth government decided that drought would no longer be covered by these natural disaster relief arrangements. The main impetus for this decision was budgetary; drought was accounting for the largest proportion of disaster relief expenditure, and there was suspicion that the Queensland state government was manipulating the scheme for electoral advantage. The commonwealth Minister for Finance claimed that the Queensland government was using the scheme as "as a sort of National Party slush fund" (Walsh, 1989).

In 1989 the commonwealth government set up the Drought Policy Review Task Force to identify policy options to encourage primary producers and other segments of rural Australia to adopt self-reliant approaches to the management of drought, consider the integration of drought policy with other relevant policy issues, and advise on priorities for commonwealth government action in minimizing the effects of drought in the rural sector (Drought Policy Review Task Force, 1990). The task force reported in 1990 and recommended against reinstating drought in the natural disaster relief arrangements. They concluded that drought was a natural part of the Australian farmer's operating environment and should be managed like any other business risk. The report recommended the establishment of a national drought policy based on principles of self-reliance and risk management, with any assistance to be provided in an adjustment context, to be based on a loans-only policy and to permit the income support needs of rural households to be addressed in more extreme situations (Drought Policy Review Task Force, 1990).

B. The National Drought Policy

Commonwealth and state ministers, through the Ministerial Council, announced a new National Drought Policy in July

1992. As recommended by the Drought Policy Review Task Force, the policy was based on principles of sustainable development, risk management, productivity growth, and structural adjustment in the farm sector. Support for productivity improvement and improved risk management was to be provided through the commonwealth government's main structural adjustment program for agriculture, the Rural Adjustment Scheme, which was being reviewed concurrently with development of the National Drought Policy.

The revised Rural Adjustment Scheme incorporated the new concept of "exceptional circumstances" under which support would be made available for farm businesses faced with a downturn for which the best manager could not be expected to prepare. Eligible events were not limited to drought. The exceptional circumstances provisions became the basis for the delivery of support during the droughts of the mid-1990s and 2002–03. Support, in the form of interest rate subsidies on commercial finance, was available only to farmers with long-term viable futures in agriculture. The rationale for this approach was that drought relief should not act as a de facto subsidy to otherwise nonviable businesses. In addition to exceptional circumstances support through the Rural Adjustment Scheme, schemes were set up to enable farmers to build financial reserves as part of their risk management, and governments made a commitment to invest in research and development, including climate research, and in education and training. The state governments agreed to phase out transaction-based subsidies such as fodder subsidies, and support was made available to help nonviable farmers leave the land. Farmers who decided to exit farming were supported with reestablishment grants and a loans-based income support scheme.

The timing of the National Drought Policy, which took effect in January 1993, could not have been more unfortunate. Parts of Queensland and New South Wales, which had been experiencing dry spells since about 1991, were settling into what was to become one of the worst droughts of the 20th century. In addition, farmers had been coping with historically high interest rates and low commodity prices. These factors combined to make the notion of preparing for drought particularly problematic. The exceptional circumstances provisions of the Rural Adjustment Scheme were triggered immediately after the scheme commenced (although ironically their first use was in response to excessive rain in parts of South Australia and Victoria), and these provisions quickly came to dominate the new scheme.

By mid-1994 the drought situation was being described as the "worst on record" (Wahlquist and Kidman, 1994) and several media organizations launched a public appeal to raise funds for drought-affected farmers. In September 1994, Prime Minister Paul Keating visited one of the worst affected areas and shortly afterward announced the establishment of a welfare-based drought relief payment scheme to help farmers meet day-to-day living expenses. Unlike assistance available through the Rural Adjustment Scheme, the drought relief payment was not limited to farmers with a long-term future in farming, but it was restricted to farmers in areas declared to be experiencing exceptional circumstances. The welfare payment was only for farm families affected by drought and was not available during other forms of exceptional circumstances.

In 1997, following the end of the drought and a change of government at the commonwealth level, a review was initiated into the operation of the National Drought Policy. The review endorsed the risk management approach of the policy but recommended some changes to its operation. At the same time the drought policy was under review, the Rural Adjustment Scheme was also reviewed and subsequently wound up being replaced by a suite of programs under the title "Agriculture—Advancing Australia" (Anderson, 1997). The new programs were not dissimilar from those they replaced and continued to be aimed at improving farm productivity and risk management. The drought relief payment was retained but extended to address a wider range of exceptional circumstances beyond drought, thus being renamed the "exceptional circumstances relief payment." In 1999, commonwealth and state ministers decided to refocus exceptional circumstances support on welfare relief and phase out the business support components that had been provided through interest rate subsidies.

In 2002 and 2003, Australia experienced widespread drought, with some regions registering the lowest rainfall on record (Bureau of Meteorology, 2002). The National Drought Policy was once again put to the test, and a number of ongoing problems with the system have once again come to the fore. First, the continuing lack of an agreed-upon definition of exceptional circumstances hampers the establishment of a stable, predictable environment within which policy makers and farmers must operate. While the trigger point at which support becomes available, and the nature of that support, remains fluid, farmers' risk management strategies will be hindered and the expectation of support is likely to generate less than optimal management decisions. The term *exceptional circumstances* was not defined in either the legislation establishing the provision or any of the accompanying explanatory material, such as ministerial speeches. Attempts have been made over the life of the National Drought Policy to develop an objective, "scientific" definition of exceptional drought, but, as is generally agreed in the international literature, drought is very difficult to define (Dracup et al., 1980; Wilhite, 2000b; Wilhite and Glantz, 1985). Second, exceptional circumstances declarations have been geographically based, resulting in what has become known as the "lines on maps" problem. Thus, farmers in arguably objectively similar circumstances are treated quite differently because of the placement of the boundary delineating exceptional circumstances areas. Because considerable government support is available to those on the "right" side of the line, this is an issue of great concern. The problem was recognized in 2001 when ministers agreed to the introduction of "buffer zones" around exceptional circumstances areas so that farmers in "reasonable proximity" to but outside the defined zones could apply for support (Agriculture and Resource Management Council of Australia and New Zealand, 2001).

The application process for assistance was also changed to allow farmers to make a prima facie case that they qualified for support. If the application was subsequently rejected, farmers would still be able to receive up to 6 months of welfare support, whereas successful applications would result in income support payments for 2 years (Truss, 2002b). Eligibility was further relaxed by the decision to extend exceptional circumstances declarations to an entire state once 80% of it qualified under the exceptional circumstances program.

During the 2002–03 drought there was evidence of some success with the risk management approach to drought preparation. In 2002, Australian farmers held approximately AU$2 billion in farm management deposits, a special scheme to help farmers build financial reserves in preparation for downturns such as drought.

C. Current Status and Future Directions

In 2004 a national roundtable was convened to consider drought policy. The roundtable considered a paper produced by an independent panel following consultations with stakeholders, and the roundtable results will be considered by government (Truss, 2003b). A number of issues need to be addressed. First, because the policy is dependent on the declaration of an exceptional circumstances drought, the process of drought declaration has become highly politicized. As is often the case in Australia, the commonwealth and state governments are from different political parties, which has created an opportunity for politicians to use drought relief to score political points (Amery, 2002; Truss, 2002a). This problem presents itself in ongoing debates about funding responsibilities for drought support as well as in relation to the second problem with the system—the definition of "exceptional cirumstances". Third, the existing system is expensive, with cost estimates of drought relief in the 2002–03 drought exceeding AU$1 billion (Truss, 2003a). A number of questions of equity are associated with this expenditure. The taxpayers who contribute to the drought support are often less wealthy over their lifetimes than the farmers—often temporarily cash poor because of drought but asset rich because of their ownership of land—who are assisted. Potential inequities also exist between farmers, particularly between those who do not qualify for support because of sound financial management and poorer managers who find themselves in difficulty.

Australia has successfully introduced a national drought policy based on a recognition of the reality of the Australian climate, and it emphasizes preparedness rather than disaster response. However, the Australian experience with policy implementation contains important lessons about the introduction of such a policy.

First, it is highly problematic to introduce a drought policy based on risk management during a severe drought event. Second, basing any drought relief on drought declarations brings up two major implementation problems: the definition of the circumstances under which support will be available and the inequities raised by geographical delineation of the eligible areas. Third, a system that relies on farmers to make a case for support rather than satisfying previously agreed-upon criteria opens the system to politicization, particularly in a federal system in which both levels of government are involved in the delivery of drought relief.

Policy makers, academics, and rural commentators generally agree that the underlying principles of the National Drought Policy are sound (see, e.g., Botterill and Fisher, 2003). It is difficult to argue against the proposition that drought is a normal feature of the Australian environment and that farmers need to manage climate risk along with other business risks they face.

Given the problems with the existing system, several alternative policy options are available. One of the most appealing approaches would see the removal of broad-brush drought declarations, to be replaced by a system that delivers support to farmers on an individual basis. This would depoliticize the policy process and ensure that support was directed where it was most needed. One mechanism for delivering this style of support would be through revenue-contingent loans similar to the Higher Education Contribution Scheme in Australia (for more detail on this proposal, see Botterill and Chapman, 2002). Under this type of arrangement, eligible farmers would access a loan only to be repaid when the farm's revenue stream returned to normal levels.

The advantages of such an approach are that it is consistent with a risk management approach to drought because the farmer draws on future good times to help him or her through current difficulties, it would not require geographically defined drought declarations, and it would be administratively simple.

Another approach could be to return exceptional drought to the natural disaster relief arrangements on the same footing as other natural disasters such as floods, cyclones, and earthquakes. A standing arrangement along these lines would greatly reduce the politicization of the process but would require an agreed-upon definition of drought and geographical delineation of areas that were eligible for support—both of which have been stumbling blocks in the current system.

A further consideration would be a reexamination of welfare support for farmers in Australia. Australian agriculture is dominated by the family farm, and for many farm families there is close integration of the farm business and the farm family. Although political leaders have been stressing for some time that farming is a business (Anderson, 1997; Crean, 1992), for many farmers the distinction between farm business and the farm family remains blurred. Australia's social welfare safety net does not cope well with the self-employed nor, since the introduction of asset testing in the 1980s, with supporting those in the community who are asset rich but income poor. To date, farm welfare support in Australia has been delivered as part of structural adjustment packages, with the objective of encouraging marginal farmers to leave the land. Although these programs have been largely unsuccessful (Botterill, 2001), governments continue to frame farm poverty as a structural adjustment issue. As noted above, the emphasis of the exceptional circumstances program in recent years has shifted to the delivery of welfare support. If the general welfare safety net were adapted to deliver short-term income relief to farmers in difficulty, there would be less pressure for special payments during drought.

The removal of drought from the natural disaster relief arrangements in Australia in 1989 signaled a major shift among policy makers from a position arguably anchored in European expectations of rainfall to a recognition of the realities of the uncertain Australian climate. The new policy environment emphasized the responsibility of farmers to manage climate risk with governments to "create the overall environment which is conducive to this whole farm planning and risk management approach" and to "act to preserve the social and physical resource base of rural Australia" in cases of severe downturn (Agricultural Council of Australia and New Zealand, 1992, p. 13). Although this new approach has general currency among members of the rural policy community, it is less clear that it has been understood or accepted by the general public, both farming and nonfarming. As Wahlquist has argued, Australia's media does not have a good record of presenting in-depth analysis of rural issues (Wahlquist, 2003), and coverage of drought is patchy and often inconsistent. This inconsistency, and the generous public response to an appeal in support of drought-affected farmers, suggests that the message that drought is a normal part of the Australian environment has not filtered through to the general community. Bushfires in Canberra and Sydney and water restrictions in urban areas in recent years have perhaps improved the broader understanding of the impact of drought.

The 2004 review of the National Drought Policy will be an important test of the Australian policy process, because 2004 will be a federal election year and a national party minister is likely to be wary of too strong a policy stance that puts further responsibility on farmers to manage for drought. Key farm groups are engaged in their own internal consultation processes in preparation for the roundtable (e.g., New South Wales Farmers Association, 2003). Issues of drought preparation, declaration processes, drought definitions, and appropriate forms of government support are all likely to be debated in detail. The policy is starting from a philosophical base that recognizes the reality of Australia's climate. The challenge is to ensure that the next review builds on this in order to achieve a sustainable and equitable drought response.

IV. DROUGHT POLICY IN SOUTH AFRICA

South Africa has a long history of living with drought. A drought during the early 1930s that coincided with the great depression made a deep impression on many policy makers. Significant droughts also occurred during the 1960s, 1980s, and early 1990s. Despite this familiarity with drought, policy makers still struggle to quantify it and to develop a stable policy framework. Drought policy falls at the interface among the numerous definitions of drought that require some quantification of intensity, duration, and geographical extent; the demand of human activities for water; and the safeguarding of the natural environment. Therefore drought policy continues to evolve, particularly with the dynamic political environment in South Africa.

South Africa is characterized by east–west degradation in rainfall, from greater than 1000 mm in the east to around 150 mm in the west. Much of the country lies above the escarpment (1000 m) and experiences a combination of frontal and convective rainfall, falling mainly during summer. The southern coast receives rainfall throughout the year and the southwestern corner is dominated by winter rainfall. The 500-mm isohyet divides the country into arable land to the east and primarily rangeland farming to the west. South Africa, receiving a little less than 500 mm as a national average, is classified as a dry country where the influence of variable rainfall cannot be underestimated.

Part of the difficulty in addressing drought in South Africa is the large proportion of the population that depends on rainfed subsistence agriculture. This sector relies heavily on the success of the rainy season to maintain adequate stocks of food. Historically, the infrastructure development and records maintenance in these areas have been neglected, which has made it difficult to monitor food status.

Traditionally, the broad definition of drought in South Africa has been seasonal rainfall less than 70% of normal (Bruwer, 1990). Using this criterion, drought has been shown to occur about 1 in 3 years in the western and northwestern regions of the country. Only 30% of the country receives 500 mm per annum or more. To further emphasize this, less than 18% of the country can be classed as arable land, of which 8% has fairly serious limitations to arable production (Schoeman et al., 2000). This underlines the vulnerability of the country to the vagaries of rainfall.

In one of the early examinations into the causes of drought, the 1923 Drought Investigation Commission concluded:

> Whether the character has altered or not, its quantity diminished, drought losses can be fully explained without presuming a deterioration in the rainfall. Your Commissioners had a vast amount of evidence placed before them from which only one conclusion can be drawn, namely, that the severe losses of the 1919 drought were caused principally by faulty veld [rangeland] and stock management. (Union of South Africa, 1923 p.5).

Subsequent investigations into various aspects of agriculture in the 1960s and 1970s reiterated this observation, indicating that no real lessons were learned, and once conditions returned to normal the policy status quo was generally maintained. Some initiatives were implemented to reduce overgrazing: conversion of cropland to grazing land in marginal areas and a revised scheme to limit assistance only to farmers who followed sound agricultural and financial practices.

In the past, state aid required magisterial districts (third-tier government) to be declared "drought disaster" status. This legal requirement necessitated a quantitative index that could be uniformly applied. This index was broadly defined as two consecutive seasons of 70% or less rainfall (Bruwer, 1990). Normal drought, for which a farmer was expected to be self-reliant, was for a period of 1 year or less. Disaster drought was defined as two consecutive seasons of below 70% of normal rainfall. A disaster drought implied that an area would qualify for state relief.

In fact, Bruwer (1990) noted that certain magisterial districts had been declared disaster drought areas for 70% of a 30-year review period, whereas some eastern portions of the country had never been declared. This indicates that the quantitative index is not optimal. In fact, because it depends on a deviation from mean annual rainfall, and the skewed distribution of annual rainfall totals in the drier areas, it favored drought declarations in the lower rainfall regions.

The National Drought Committee (NDC) consisted of representatives of farmers' organizations, the Soil Conservation Advisory Board, the financial sector (i.e., banking, agricultural credit organizations, and Department of Agricultural Economics and Marketing), and the agricultural community. This committee scrutinized applications concerning disaster drought status and then advised the Minister of Agriculture regarding these applications. On the local level, a district drought committee was formed under the chair of the local magistrate. This committee examined all local applications and submitted these to the NDC using the prescribed format. Declaration and revocation of drought-stricken areas were evaluated considering the following five criteria:

1. Rainfall over at least three seasons
2. Veld (rangeland) condition
3. Availability of water (for stock)
4. Stock condition or deaths
5. Availability and volume of fodder to be purchased

Drought assistance schemes were aimed at maintaining a nucleus herd or stock for reestablishment after the drought was over. A phased system of drought assistance was developed. The first level consisted of rebates on transport costs, followed by loans and finally subsidies at increasing rates as the drought continued. The assistance scheme served to protect natural resources and provide for livestock farmers during a disaster drought. The maintenance of a healthy and viable nucleus herd was not to be at the expense of the natural resources or to the detriment of a farmer's financial position. However, by the mid 1980s it was clear the policy had failed to protect natural resources as envisaged.

The government acknowledged that the drought assistance schemes contributed to sustaining selected agricultural production and communities. However, they struggled to define clear onset and shut-down phases of the drought assistance, and at times, assistance was out of phase with environmental conditions.

Many farmers overestimated the condition and potential of their agricultural resources. Farmers who overexploited their resources also benefited from the drought assistance schemes. This was clearly unacceptable. The scheme did not encourage a proactive or risk management approach. During their visit to South Africa, White and O'Meagher (1999) observed that inappropriate policy and incentives had led to inappropriate management, which resulted in nonviable agriculture and degradation of scarce natural resources. As noted above, Australian drought policy places considerable emphasis on encouraging primary producers to adopt self-reliant approaches to cope with drought and farm management.

For stock farmers, the government provided assistance in time of drought for the movement of stock or fodder and availability of loans. The definition of drought resulted in certain areas being under a disaster drought declaration for >50% of the time between 1956 and 1986 (Smith, 1994). Incidental observation of pre-1990 drought policy for stock farmers noted that periods of drought declaration were excessive, relating more to overstocking than to climate. The effects on land degradation were serious, and government assistance, while substantial, could be interpreted as exacerbating the problem rather than reducing it (Smith, 1994). The challenge of determining when intervention should occur and in what form has occupied experts in many countries. Until the 1990s, drought policy in South Africa was directed primarily at stock farmers (Walters, 1993). Stock farming was considered best adapted to the highly variable rainfall conditions in these areas. However, assistance tended to favor the poorer managers and climatically marginal area (Smith, 1993).

Bruwer (1990) pointed out that most drought countermeasures were reactive in nature. The development of new policy in the early 1990s required greater emphasis on a proactive approach. Bruwer also pointed out that drought effects are largely human induced. This signified an important turnaround in the approach of government policy regarding state drought intervention. Rangeland specialists identified farming areas that were overstocked by >50%. Research in the Free State province demonstrated how overstocking increases the length of fodder shortage periods and the probability of such shortages.

The government concluded that if no remedial actions were taken, land degradation could encroach on a significant proportion of the country (Bruwer, 1990). Expenditure on drought and flood relief was increasing significantly, from $150 million during 1984–85 to $330 million in 1992–93 (Monnik, 1997). Was this expenditure justified in achieving the government's objectives? With the continued degradation of natural agricultural resources, government was provided with a strong motivation to review its approach in providing financial and other relief. As put by Tyson (1988 p.17), who emphasized the need for a different paradigm: "All future planning must be predicated on the assumption that it is a land of drought rather than a land of plentiful rain."

During the 1980s, drought stakeholders in South Africa were captivated by a sense of anticipation. Research by Tyson (1986) concerning rainfall patterns and cycles of wet and dry spells on a decadal scale led to the successful prediction of the drought during the early 1980s. Van Heerden et al. (1988) and many others internationally investigated the feasibility of providing seasonal rainfall anomaly predictions. The first-ever "official" long-term prediction was attempted for the 1986–87 summer season. Because of possible misunderstanding, the forecast was submitted personally to interested parties and not released to the media (Van Heerden, 1990). Tyson's (1986) research on 33 widely distributed rainfall sites across the summer rainfall region of South Africa showed a clear oscillatory pattern in rainfall, producing 9-year spells of alternating generally dry and wet conditions.

Parallel to the progress in seasonal forecasts were developments in satellite-based operational monitoring systems such as the National Oceanic and Atmospheric Administration's or NOAA's Advanced Very High Resolution Radiometer, or AVHRR. Coupled with the availability of crop and rangeland models and rainfall deciles, there was a sense of anticipation that effective monitoring and forecasting were coming together. Dr. J Serfontein, the then deputy director-general of environmental affairs, speaking at the Southern African Regional Commission for the Conservation and Utilization of the Soil (SARCCUS) Workshop on drought, stressed the benefit such information could be to government for long-range planning (Serfontein, 1990).

Du Pisani (1990) summarized some of the objective techniques used to monitor drought in South Africa. These included utilization of water balance models, rainfall deciles, crop models, and remote sensing techniques. He highlighted the difficulty in determining drought severity because of the interaction of intensity, geographical extent, duration, and water resource requirements.

During the early 1990s, drought policy was changed to place greater obligations on farmers to reciprocate for state aid. Farmers were asked to commit to practices aimed at promoting resource conservation and the long-term sustainability of economic production (Walters, 1993). For example, only farmers who submitted their stocking rates (stock units per hectare) quarterly to their local Department of Agriculture office, and who reduced stock on drought warnings, were eligible for state aid. Although this policy led to improvements in the management of drought assistance for stock farmers, it continued the institutionalized neglect for the protection of the rural poor from threats posed by insufficient water, food, and employment (Walters, 1993). In addition, the definition of disaster drought conditions continued to favor the western portions of the country, where the coefficient of variation of rainfall was greater.

One of the keys to the new drought policy (post-1990) was the recognition of grazing capacity zones (Smith, 1993). Five grazing capacity classes were defined across the country. Drought assistance was limited to farmers who remained within the prescribed capacities. Farmers had to maintain records of stock numbers and report on a quarterly basis. Drought declarations were proposed by local drought committees and supervised by the National Drought committee. The director-general made the final decision, basing the declaration on the same five factors used by the NDC that were mentioned previously.

One of the central aims of the system was to protect natural resources, which had been a largely unrealized aspiration of the original policy. By elevating this within the policy framework, the goal was to take better care of natural resources in South Africa. In fact, since the 1960s, there has been little incentive for farmers to refrain from cultivating marginal agricultural land (Vogel, 1994). Inappropriate subsidies and a high level of mechanization and fertilization resulted in deterioration of farmland in many instances.

The approach during the early 1990s attempted to develop a longer term view with a greater emphasis on risk management. Part of the motivation was to address the apparent escalation in the cost of implementing the existing schemes. The scheme as developed in 1990 aimed to:

- Provide financial assistance to farmers
- Ensure agricultural resources are protected
- Encourage farmers to apply optimal resource utilization
- Contribute to the maintenance of a nucleus breeding herd

These aims did not differ much from the original aims, although the scheme did attempt to address some of the concerns that had become evident over the past years.

In 1990, the government argued that, because South Africa is an arid country, the state has a moral obligation and responsibility to assist people during times of hardship and prevent long-term disruption to communities and infrastructure (Bruwer, 1990). Although the systems seemed to address the situation among commercial stock farmers, subsistence farmers, crop farmers, and other significant stakeholders were overlooked. In contrast to this, the then Minister of Agriculture just 7 years later noted that drought aid encouraged bad practice, was inequitable in the past, and created expectations that government would bail out farmers in all disasters; he also noted that a prolonged drought would affect everyone in the country (Hanekom, 1997).

During the early 1990s, the National Consultative Forum on Drought was formed. This forum, comprising members of government and nongovernmental organizations, sought to raise the profile of all communities that were being affected by drought. This allowed for drought assistance to be provided to a broader range of the population (Vogel, 1994).

The White Paper on Agriculture (Ministry of Agriculture, 1995 p. 7) stated that: "Drought will be recognised as a normal phenomenon in the agricultural sector and it will be accommodated as such in farming and agricultural financing systems."

The new democratic government's stand on drought assistance was still to be tested during a "real" drought. The farming community, which had generally benefited in the past from the majority of government assistance, expressed concern. The acknowledgment that drought was part of the normal environment caused them to reevaluate their practices.

Thus as media attention focused on the El Niño/Southern Oscillation (ENSO) phenomenon during 1997, when there was an expectation of a large El Niño event, response from the private sector was noticeable. It was reported that McCarthy Motor Holdings sold no light delivery vehicles from their Hoopstad agency after the first El Niño press release (September 1997). In addition, tractor sales for October 1997 were 20% lower and sales of haymaking equipment increased by 50% over the corresponding period the previous year (Redelinghuys, 1997).

The impact of this ENSO event on South African rainfall did not materialize as predicted, which caused many people to lose faith in these forecasts. However, economists observed that the financial discipline exerted by farmers resulted in them being in a much healthier state at the end of the season than if they had ignored the forecast from the beginning. This indicated that commercial farming can benefit from a greater appreciation of drought risk. It may also contribute to developing some degree of robustness to drought.

The discussion paper on agricultural policy in South Africa provided more detail concerning the new government's view on drought assistance. The authors recognized that past policies had weakened the farmers' resolve to adopt risk-coping strategies. Expenditure of $1.2 billion to write off and consolidate farmer debt during 1992–93 was identified as unsustainable. The government set itself on a course to provide other options, besides relief, to help farmers cope with drought (Ministry of Agriculture, 1998).

More recently, the government has placed more emphasis on risk management. During 2002, an agricultural risk insurance bill was developed. The purpose of the bill was to enhance the income of those farmers and producers most vulnerable to losses of agricultural crops and livestock due to natural disaster, including drought. In addition, a drought management strategy was under development during 2003 and 2004. This document is eagerly anticipated to provide greater detail in line with the policy guidelines.

Williams (2000) pointed out that the recent advances in long-lead forecasting provide the opportunity to focus more on managing climatic variability instead of being the passive victim of an "unexpected" drought. South Africa needs to maintain its investment in meteorological research and communication to the public, and to encourage links with the global meteorological community.

A challenge remains for the South African government: to maintain a policy balance between encouraging a risk management approach for large agricultural enterprises and providing a safety net for the resource-limited sectors of the population.

V. MOVING FROM CRISIS TO RISK MANAGEMENT: CREEPING TOWARD A NATIONAL DROUGHT POLICY FOR THE UNITED STATES

Drought is a normal part of the climate for virtually all portions of the United States; it is a recurring, inevitable feature of climate that results in serious economic, environmental, and social impacts. The Federal Emergency Management Agency (FEMA) estimates average annual losses because of drought in the United States to be $6–8 billion, more than for any other natural hazard (FEMA, 1995). Yet the United States is ill prepared to effectively deal with the consequences of drought. Historically, the U.S. approach to drought management has been to react to the impacts of drought by offering relief to the affected area. These emergency response programs can best be characterized as too little and too late. More important, as noted in this chapter for Australia and South Africa, drought relief does little if anything to reduce the vulnerability of the affected area to future drought events. In fact, there is considerable evidence that providing relief actually increases vulnerability to future events by increasing dependence on government and encouraging resource managers to maintain the very resource management methods that may be placing the individual, industry, utility, or community at risk. Improving drought management requires a new paradigm, one that encourages preparedness and mitigation through the application of the principles of risk management.

Drought conditions are not limited to the western United States—although they occur more frequently in this region and are usually longer in duration than those that occur in the east. The droughts of 1998–2002 demonstrated the vulnerability of the eastern states to severe and extended periods of precipitation deficits. Wherever it occurs, severe drought can result in enormous economic and environmental impacts as well as personal hardship. However, because the incidence of drought is lower in the east, this region is generally less prepared to mitigate and respond to its effects. The west is currently better equipped to manage water supplies during extended periods of water shortage because of large investments in water storage and transmission facilities, more advanced water conservation measures, irrigation, and other measures that improve resiliency.

State-level drought planning has increased significantly during the past two decades (Wilhite, 1997a). In 1982, only 3 states had drought plans in place. By 2004, 36 states had developed plans and 4 states were at various stages of plan development (*http://drought.unl.edu/mitigate/status.htm*). The basic goal of state drought plans should be to improve the effectiveness of preparedness and response efforts by enhancing monitoring and early warning, risk and impact assessment, and mitigation and response. Plans should also contain provisions to improve coordination within agencies of state government and between local and federal government. Initially, state drought plans largely focused on response; today the trend is for states to place greater emphasis on mitigation as the fundamental element. Several states have recently revised their drought response plans to further emphasize mitigation (e.g., Montana, Nebraska, Colorado). Other states that previously did not have a drought plan have recently developed plans that place more emphasis on mitigation (e.g., New Mexico, Texas, Georgia, Hawaii). Arizona is currently developing a drought mitigation plan. As states gain more experience with drought planning and mitigation actions, the trend toward mitigation is expected to continue. In addition, drought planning must be considered an ongoing process rather than a discrete event. Moving from response planning to mitigation planning represents a continuum. Even the most advanced state drought planning efforts have moved only partially along that continuum.

The growth in the number of states with drought plans suggests an increased concern at that level about the potential impacts of extended water shortages and an attempt to address those concerns through planning. Initially, states were slow to develop drought plans because the planning process was unfamiliar. With the development of drought planning models (see Chapter 5) and the availability of a greater number of drought plans for comparison, drought planning has become a less mysterious process for states (Wilhite, 2000a). As states initiate the planning process, one of their first actions is to study the drought plans of other states to compare methodology and organizational structure.

The rapid adoption of drought plans by states is also a clear indication of their benefits. Drought plans provide the framework for improved coordination within and between levels of government. Early warning and monitoring systems are more comprehensive and integrated, and the delivery of this information to decision makers at all levels is enhanced. Many states are now making full use of the Internet to disseminate information to a diverse set of users and decision makers. Through drought plans, the risks associated with drought can be better defined and addressed with proactive mitigation and response programs. The drought planning process also provides the opportunity to involve the numerous stakeholders early and often in plan development, thus increasing the probability that conflicts between water users will be reduced during times of shortage. All of these actions can help to improve public awareness of the importance of water management and the value of protecting limited water resources.

With tremendous advances in drought planning at the state level in recent years, it is not surprising that states have been extremely frustrated and dissatisfied with the lack of progress at the federal level. The lack of federal leadership and coordination quickly became an issue after a string of consecutive drought years beginning in 1996. This resulted in a series of policy initiatives that have put the United States on course to develop a national drought policy.

Calls for action on drought policy and plan development in the United States date back to at least the late 1970s. The growing concern has resulted primarily from the inability of the federal government to adequately address the spiraling impacts associated with drought through the traditional reactive, crisis management approach. This approach has relied on ad hoc inter-agency committees that are quickly disbanded following termination of the drought event. The lessons (i.e., successes and failures) of these response efforts are forgotten and the failures are subsequently repeated with the next event. Calls for action include recommendations from the Western Governors' Policy Office (1978), General Accounting Office (1979), National Academy of Sciences (1986), Great Lakes Commission (1990), Interstate Council on Water Policy (1991), Environmental Protection Agency (Smith and Tirpak, 1989), American Meteorological Society (1997), Office of Technology Assessment (1993), Federal Emergency Management Agency (1996), Western Governors' Association (1996), and Western Water Policy Review Advisory Commission (1998).

The most recent of these calls for action are worthy of further discussion. In response to the severe impacts of drought in 1996, FEMA was directed to chair a multi-state drought task force to address the drought situation in the Southwest and the southern Great Plains states (FEMA, 1996). The purpose of the task force was to coordinate federal response to drought-related problems in the stricken region by identifying needs, applicable programs, and program barriers. The task force was also directed to suggest ways to improve drought management through both short- and long-term national actions. The final report of this task force contained several important long-term recommendations. First, the task force called for the development of a national drought policy based on the philosophy of cooperation with state and local stakeholders. It recommended that this policy include a national climate and drought monitoring system to provide early warning to federal, state, and local officials of the onset and severity of drought. Second, it suggested that a regional forum be created to assess regional needs and resources, identify critical areas and interests, provide reliable and timely information, and coordinate state actions. Third, FEMA was asked to include drought as one of the natural hazards addressed in the National Mitigation Strategy, given the substantial costs associated with its occurrence and the numerous opportunities available to mitigate its effects. Fourth, states strongly requested that a single federal agency be appointed to coordinate drought preparedness and response.

Another important initiative resulting from the 1996 drought was the development of a drought task force under the leadership of the Western Governors' Association (WGA). This task force, formed in June 1996, emphasized the importance of a comprehensive, integrated drought response. The WGA Drought Task Force's report made several important recommendations (WGA, 1996). First, it recommended development of a national drought policy or framework to integrate actions and responsibilities among all levels of government and emphasize preparedness, response, and mitigation measures. Second, it encouraged states to develop drought preparedness plans that include early warning, triggers, and short- and long-term planning and mitigation measures. Third, it called for creation of a regional drought coordinating council to develop sustainable policy, monitor drought conditions, assess state-level responses, identify impacts and issues for resolution, and work in partnership with the federal government to address drought-related needs. Fourth, the report called for establishment of a federal interagency coordinating group with a designated lead agency for drought coordination with states and regional agencies.

A number of important policy initiatives have resulted from the FEMA and WGA reports. A memorandum of understanding (MOU) was signed in early 1997 between the WGA and several federal agencies. This MOU called for a partnership between federal, state, local, and tribal governments to reduce drought impacts in the western United States. The MOU resulted in the following actions: (1) the Western Drought Coordination Council (WDCC) was formed to address the recommendations of the western governors; (2) the U.S. Department of Agriculture (USDA) was designated as the lead federal agency for drought, to carry out the objectives of the MOU; and (3) the USDA established a federal inter-agency drought coordinating group.

Another initiative of considerable relevance was the reexamination of western water policy by the Western Water Policy Review Advisory Commission (1998). This commission was created by passage of the Western Water Policy Review Act of 1992. One of the commission's reports summarized recommendations from recent studies on drought management that should be incorporated in future attempts to integrate drought management and water policy in the West (Wilhite, 1997b). The consensus from the reports reviewed in this study emphasized the need for a national drought policy and plan, a national climate monitoring system in support of that policy, and the development of state drought mitigation plans. Although impacts of drought occur mainly at the local, state, and regional level, this study concluded that it was imperative for the federal government to provide the leadership necessary to improve the way the nation prepares for and responds to drought.

The National Drought Policy Act of 1998 (PL 105–199) was introduced in Congress as a direct result of the 1996 drought and the initiatives referred to previously. This bill created the National Drought Policy Commission (NDPC) to "provide advice and recommendations on creation of an integrated, coordinated Federal policy designed to prepare for and respond to serious drought emergencies." The NDPC's report, submitted to Congress and the president in May 2000, recommended that the United States establish a national drought policy emphasizing preparedness (NDPC, 2000). The goals of this policy would be to:

1. Incorporate planning, implementation of plans and proactive mitigation measures, risk management, resource stewardship, environmental considerations, and public education as key elements of an effective national drought policy
2. Improve collaboration among scientists and managers to enhance observation networks, monitoring, prediction, information delivery, and applied research and to foster public understanding of and preparedness for drought
3. Develop and incorporate comprehensive insurance and financial strategies into drought preparedness plans
4. Maintain a safety net of emergency relief that emphasizes sound stewardship of natural resources and self-help
5. Coordinate drought programs and resources effectively, efficiently, and in a customer-oriented manner

The NDPC further suggested creation of a long-term, continuing National Drought Council composed of federal and nonfederal members to implement the recommendations of the NDPC. It advised Congress to designate the Secretary of Agriculture as the co-chair of the council, with a nonfederal co-chair to be elected by the nonfederal council members. An interim National Drought Council was established by the Secretary of Agriculture following submission of the NDPC report, pending action on a permanent council by the U.S. Congress.

In July 2003, the National Drought Preparedness Act was introduced in the U.S. Congress. The purpose of this bill is "to improve national drought preparedness, mitigation, and response efforts." The bill authorizes creation of a National Drought Council within the Office of the Secretary of Agriculture. Membership on the council would be composed of both federal and nonfederal persons. The council would assist in coordinating drought preparedness activities between the federal government and state, local, and tribal governments. A National Office of Drought Preparedness would be created within the USDA to provide assistance to the council. The council is directed by the bill to develop a "comprehensive National Drought Policy Action Plan that

- delineates and integrates responsibilities for activities relating to drought (including drought preparedness, mitigation, research, risk management, training, and emergency relief) among Federal agencies; and
- ensures that those activities are coordinated with the activities of the States, local governments, Indian tribes, and neighboring countries; and
- is integrated with drought management programs of the States, Indian tribes, local governments, watershed groups, and private entities; and
- avoids duplicating Federal, State, tribal, local, watershed, and private drought preparedness and monitoring programs in existence."

This bill also stresses improvement of the national integrated drought monitoring system by enhancing monitoring and climate and water supply forecasting efforts, funding specific research activities, and developing an effective drought information delivery system to improve the flow of information to decision makers at all levels of government and to the private sector. A preliminary study to assess gaps in the current drought monitoring network and compile a prototype of a more comprehensive, integrated national drought information system was recently completed with support from the NOAA, under the leadership of the WGA (2004).

Actions taken since 1996 to improve drought management in the United States have had little effect to date—especially at the federal level, as verified by the federal response to drought conditions in 2000–2003. Instead, states have continued to be the most progressive, a trend that began in the early to mid-1980s. Thirty-six states have drought plans and another four states are at various stages of plan development, most with a focus on mitigation. Other states have made substantial progress in drought plan revision, again emphasizing mitigation. Federal agencies are now speaking the new language of drought management, and phrases like "improved coordination and cooperation," "increased emphasis on mitigation and preparedness," and "building nonfederal/federal partnerships" have become commonplace. However, the existing institutional inertia of federal emergency response programs and the expectations of the recipients of those assistance programs encourage drought management to remain in a reactive, crisis management mode. The mentality of most state and federal government agencies clearly remains response oriented. Whether federal and state policy makers clearly understand the scope of the changes that will be required to invoke the new paradigm of risk management in the United States is not apparent at this time. When drought conditions exist, especially in election years, drought relief is one method members of Congress use to send money home to their constituents. The true test of whether we are making progress will be if the Congress passes the National Drought Preparedness Act and the USDA rapidly implements its various components. State governments and special interest groups must show their support for this bill, both when Congress is deliberating it and following its passage. Hopefully, this bill will provide the authority necessary to direct federal agencies to modify existing policies and programs to emphasize mitigation and preparedness, thus effectively shifting funding from crisis to risk management and implementing the new paradigm.

Only time will determine the dedication of the nation to this new approach to drought management. A continuation of widespread, severe drought in the next few years would certainly engender greater support for this new paradigm and help the United States continue down the path to risk management. The political will to change the way we manage drought appears to be genuine but may evaporate quickly if a series of wet years occurs. Changing the momentum of the past is a difficult obstacle to overcome. It is critical for the scientific community and the public to hold policy makers to this commitment.

VI. SUMMARY

Australia, South Africa, and the United States are extremely drought-prone nations with a longstanding history of government intervention in the form of drought relief. Drought impacts are substantial, and each government has addressed drought primarily through the crisis management approach. This approach has proved to be unsuccessful. Australia was the first of the three countries to move toward a national drought policy that emphasized a more risk-based management approach, focused on improving self-reliance and minimizing the need for government intervention during and in the post-drought period. South Africa and the United States have each followed a similar course of action and are at various stages in the development of a national drought policy. The lessons learned in each of these cases can be instructive to both developed and developing countries seeking a more proactive approach to drought management and improved levels of drought preparedness.

REFERENCES

Agricultural Council of Australia and New Zealand (ACANZ). Record and Resolutions: 138th Meeting, Mackay, 24 July 1992 (p. 13). Commonwealth of Australia, 1992.

Agriculture and Resource Management Council of Australia and New Zealand (ARMCANZ). Record and Resolutions: Twenty first Meeting, Darwin, 17 August 2001 (p. 33). Canberra, Commonwealth of Australia, 2001.

American Meteorological Society. *AMS Statement on Meteorological Drought* (pp. 847–849). Boston, MA: AMS, 1997.

Amery R, the Hon MP. Howard Government Paying Little and Wanting to Pay Less. Media Release by the NSW Minister for Agriculture, 11 October 2002.

Anderson J, the Hon MP. Federal Government Gives Farm Sector "AAA" Rating. Media Release by Minister for Primary Industries and Energy, 14 September 1997.

Bigge JT. Report on Agriculture and Trade in NSW. Adelaide: Libraries Board of South Australia (p. 92), [1823] 1966.

Botterill LC. Rural policy assumptions and policy failure: The case of the re-establishment grant. *Australian Journal of Public Administration* 60:13–20, 2001.

Botterill L, B Chapman. Developing Equitable and Affordable Government Responses to Drought in Australia. Centre for Economic Policy Research, Discussion Paper No. 455, 2002.

Botterill LC, M Fisher, eds. *Beyond Drought: People, Policy and Perspectives*. Melbourne: CSIRO Publishing, 2003.

Bruwer JJ. Drought policy in the Republic of South Africa. In: AL du Pisani, ed. *Proceedings of the SARCCUS Workshop on Drought* (pp. 23–38). Pretoria, South Africa, 1990.

Bureau of Meteorology (BOM). *Rainfall Deficiencies Worsen Following Dry October.* Melbourne: BOM, 2002.

Crean S, the Hon MP. Rural Adjustment Bill 1992: Second Reading Speech. House of Representatives Hansard, Parliament of Australia, 1992.

J Dracup, KS Lee, EG Paulson. On the definition of droughts. *Water Resources Research* 16:297–302, 1980.

Drought Policy Review Task Force (DPRTF). *National Drought Policy*, Vol. 1. Canberra: AGPS, 1990.

Du Pisani AL. Drought detection, monitoring and early warning. In: AL du Pisani, ed. *Proceedings of the SARCCUS Workshop on Drought* (pp. 6–9). Pretoria, South Africa, 1990.

Federal Emergency Management Agency. *National Mitigation Strategy* (p. 2). Washington, D.C.: FEMA, 1995.

Federal Emergency Management Agency. *Drought of 1996: Multi-State Drought Task Force Findings*. Washington, D.C.: FEMA, 1996.

Flannery T. *The Future Eaters* (p. 81). Sydney: Reed New Holland, 1994.

General Accounting Office. *Federal Responses to the 1976–77 Drought: What Should Be Done Next*. Washington, D.C.: GAO, 1979.

Great Lakes Commission. *A Guidebook to Drought Planning, Management and Water Level Changes in the Great Lakes*. Ann Arbor, MI, 1990.

Hanekom D. Agriculture's Response to El Niño and Drought Subsidies to Farmers. Media statement from Derek Hanekom, Minister for Agriculture and Land Affairs, 18 September 1997.

Interstate Council on Water Policy. *Statement of Policy 1991–92*. Washington, D.C., 1991.

Ministry of Agriculture. White Paper on Agriculture, Ministry for Agriculture and Land Affairs, Pretoria, South Africa (*http://www.gov.za/whitepaper/1995/agriculture.htm*), 1995.

Ministry of Agriculture. Agricultural policy in South Africa. Ministry for Agriculture and Land Affairs, Pretoria, South Africa (*http://www.nda.agric.za/docs/policy98.htm*), 1998.

Monnik KA. Agricultural management issues related to drought. Unpublished input paper for Task Team on Drought and other Agricultural Disasters, Ministry of Agriculture and Land Affairs, 1997.

National Academy of Sciences. *The National Climate Program: Early Achievements and Future Directions*. Washington, D.C.: NAS, 1986.

National Drought Policy Commission. *Preparing for Drought in the 21st Century*. Washington, D.C.: NDPC, 2000.

New South Wales Farmers Association. Preparing for Drought: The Key to Sustainability. Discussion Paper, September 2003.

Office of Technology Assessment, U.S. Congress. *Preparing for an Uncertain Climate* (OTA-0-567, Volume 1, pp. 250–257). Washington, D.C.: OTA, 1993.

O'Meagher B, M Stafford Smith, DH White. Approaches to integrate drought risk management. In: DA Wilhite, ed. *Drought: A Global Assessment* (Volume 2). London: Routledge, 2000.

Redelinghuys J. The Economic Implications of El Niño. National Workshop on the El Niño Phenomenon, National Department of Agriculture, Pretoria, South Africa, 3 December 1997.

Schoeman JL, M van der Walt, KA Monnik, A Thackrah, J Malherbe, RE Le Roux. Development and application of a land capability classification system for South Africa. ARC-Institute for Soil, Climate and Water report, GW/A/2000/57 (*http://www.agis.agric.za/agisweb/IDfda5eaaad3133a/$WEB_HOME?MIval=land_capability.html*), 2000.

Serfontein J. Opening address. In: AL du Pisani, ed. *Proceedings of the SARCCUS Workshop on Drought* (p. vi). Pretoria, South Africa, 1990.

Smith DI. Drought policy and sustainability: Lessons from South Africa. *Search* 24(10):292–295, 1993.

Smith DI. Drought and drought policy in the Republic of South Africa. In: KP Bryceson, DH White, eds. *Proceedings of a Workshop on Drought and Decision Support. Bureau of Resource Sciences Proceedings* (pp. 27–31). Canberra, Australia: Bureau of Resource Sciences, 1994.

Smith JB, D Tirpak (eds.). *The Potential Effects of Global Climate Change on the United States.* Washington, D.C.: EPA-230-05-89-050, 1989.

Truss W, the Hon MP. Commonwealth to Push States/Territories to Put Drought-Stricken Farmers First. Media Release by the Federal Minister for Agriculture, Fisheries and Forestry, 2 October 2002a.

Truss W, the Hon MP. Early Help for Drought-Affected Northwest NSW Farmers. Media Release by the Federal Minister for Agriculture, Fisheries and Forestry, 19 September 2002b.

Truss W, the Hon MP. $1 Billion Commitment to Help Farmers in Drought. Media Release by the Federal Minister for Agriculture, Fisheries and Forestry, 15 August 2003a.

Truss W, the Hon MP. Primary Industries Ministerial Council Meeting Communique. Media Release by Federal Minister for Agriculture, Fisheries and Forestry PIMC 4/03, 3 October 2003b.

Tyson PD. *Climate Change and Variability in South Africa* (p. 200). Cape Town: University Press, 1986.

Tyson PD. The Ever-Changing Climate of Southern Africa. Fertilizer Society of South Africa Journal 1:13–18, 1988.

Union of South Africa. *Final Report of the Drought Investigation Commission* (p. 5). CapeTimes Limited, Government printers, 1923.

Van Heerden J. Drought prediction: Some early results in southern Africa. In: AL du Pisani, ed. *Proceedings of the SARCCUS Workshop on Drought* (pp. 1–5). Pretoria, South Africa, 1990.

Van Heerden J, DE Terblanche, GC Schulze. The Southern Oscillation and South African summer rainfall. *Journal of Climatology* 8(6):577–597, 1988.

Vogel C. South Africa. In: MH Glantz, ed. *Drought Follows the Plow* (pp. 151–170). London: Cambridge University Press, 1994.

Wahlquist A. Media representations and public perceptions of drought. In: LC Botterill, M Fisher, eds. *Beyond Drought: People, Policy and Perspectives.* Melbourne: CSIRO Publishing, 2003.

Wahlquist A, M Kidman. Drought now worst in history. *Sydney Morning Herald*, 31 December 1994.

Walsh P, Senator the Hon. Question Without Notice: Natural Disaster Relief. Senate Hansard, Parliament of Australia, p. 189, 1 March 1989.

Walters MC. Present State Drought Policy in the RSA and Possible Areas of Adaptation. Presented at "Planning for Drought as a Natural Phenomenon," Mmabatho, South Africa, 28 January 1993.

Western Governors' Association. Creating a Drought Early Warning System for the 21st Century. The National Integrated Drought Information System. Denver, CO, 2004.

Western Governors' Association. Drought Response Action Plan. Denver, CO, 1996.

Western Governors' Policy Office. *Managing Resource Scarcity: Lessons from the Mid-Seventies Drought*. Denver, CO: Institute for Policy Research, 1978.

Western Water Policy Review Act of 1992. 102 U.S.C. §102–575, 1992.

Western Water Policy Review Advisory Commission. *Water in the West: Challenge for the Next Century* (pp. 5–10). Washington, D.C.: National Technical Information Service, 1998.

White DH, B O'Meagher. Proceedings of a 1999 conference on Integrating drought policy, assessment and management. In: Proceedings of the International Conference on Integrated Drought Management: Lessons for Sub-Saharan Africa. 20–22 September 1999, Pretoria, South Africa, IHP-V Technical documents in hydrology, 35:365–377, 1999.

Wilhite DA. State actions to mitigate drought: Lessons learned. *Journal of the American Water Resources Association* 33(5):961–968, 1997a.

Wilhite DA. *Improving Drought Management in the West*. Washington, D.C.: National Technical Information Service, 1997b.

Wilhite DA. Planning for drought: Moving from crisis to risk management. *Journal of the American Water Resources Association* 36(4):697–710, 2000a.

Wilhite DA. Drought as a natural hazard: Concepts and definitions. In: DA Wilhite, ed. *Drought: A Global Assessment* (Volume I, pp. 3–18). London: Routledge, 2000b.

Wilhite DA, MH Glantz. Understanding the drought phenomenon: The role of definitions. *Water International* 10:111–120, 1985.

Williams J. Drought risk management in southern Africa. In: DA Wilhite, ed. *Drought: A Global Assessment* (Volume II, pp. 168–177). London: Routledge, 2000.

7

Managing Demand: Water Conservation as a Drought Mitigation Tool

AMY VICKERS

CONTENTS

I. INTRODUCTION: A NEW *ERA* OF WATER SCARCITY OR AN OLD *ERROR* OF WATER WASTE?

The discovery from tree rings of ancient drought cycles, the emergence of centuries-old shipwrecks on drying riverbeds, and the forecasts of unruly climate change and variability can easily stir fear for our water future—in both scientist and citizen alike. Yet such conditions need not be predictors of our water fate.

Exactly how the water demands of the 21st century's growing population will be met is, indeed, a formidable challenge. Half of the world's 6 billion people now live in urban environments—projected to increase to 60% by 2030—and the majority of the globe's 16 mega-cities (10 million or more residents) reside in regions confronting mild to severe water stress, according to the United Nations (2003). Between 1950 and 2000, the world's population more than doubled (United Nations, 2002), and its water demands roughly tripled (Postel and Vickers, 2004). From 2000 to 2050, global population is projected to grow 45%, reaching nearly 9 billion people (United Nations, 2002). Clearly, the world's water demands are increasing, but nature's present—and future—water budget remains largely fixed at the limits of its primordial creation.

From where and at what cost future water supplies will be derived remains an unanswered and troubling question for many public officials and water managers. With falling groundwater tables and approximately 800,000 dams now altering natural river flows worldwide—more than 75% of the river systems in the United States, Canada, Europe, and the former Soviet Union are already diverted by dams—much of the developed world's freshwater sources have already been tapped (Postel and Richter, 2003). Signs of water stress are apparent in the receding levels of some of the world's largest and most prized bodies of fresh water: Lake Mead in Nevada, the largest human-made reservoir in the United States (Ritter, 2003); Lake Chapala, the largest freshwater body in Mexico (Carlton, 2003); and the Aral Sea in Central Asia, once the world's fourth largest lake and now a mere third of its original volume (Postel and Richter, 2003). The levels of Lake Chapala are dropping because of development and outmoded irrigation techniques used by the arid region's farmers. Cyclical droughts in the region have been aggravated by rapid population growth. That, along with declining home values for U.S. and Canadian retirees, is putting in peril the $200 million in annual revenues provided to that poor region by expatriates. The lake also is becoming a dead zone for marine life, with several fish species practically wiped out. "Time is awfully close to running out," says Dr. Woen Lind, a Baylor University biology professor who has studied Lake Chapala (Carlton, 2003).

After more than a century of water supply development and accompanying exploitation of the natural ecosystems on which water systems depend, the goal of quenching humanity's thirst for more water seems as elusive as ever. The severity and cost of the world's droughts and chronic water supply problems are worsening, arguably leading to a global water crisis. Yet, on every continent and in nearly every water system facing drought or long-term water shortage, there exists a glaring if not nagging antidote: the elimination of water waste:

> [I]t is evident that there must be a great amount of water wasted in many cities. Millions of dollars are being spent by many of our larger cities to so increase their supply that two thirds of it may be wasted. This waste is either intentional, careless, or through ignorance. (Folwell, 1900, p. 41)

> We need ... to reduce leakage, especially in the many cities where water losses are an astonishing 40 per cent or more of total water supply. (Annan, 2002)

Water waste—from leaking, neglected underground pipes to green lawns in deserts, and the application of archaic flooding methods to grow food crops—is so prevalent that it is typically considered normal if not inevitable. But is this a reasonable assumption, one that should continue to guide drought response and water management today? To be sure, all water systems will have some leaks, the human experience relies on water for its functional value as well as its aesthetic and inspirational qualities, and beneficial reuse is a component of some irrigation losses. But to what extent have we defined our *true water needs* in contrast to our *water wants, demands, and follies*? If Singapore, Copenhagen, Denmark, and Fukuoka, Japan, are able to minimize their total unaccounted-for water (UFW) losses to 5% or less, how efficiently is water used in Jordan and in Taipei, Taiwan, and Johannesburg, South Africa, that more than 40% is lost to leakage and unexplained uses? (Postel and Vickers, 2004) Does a resident

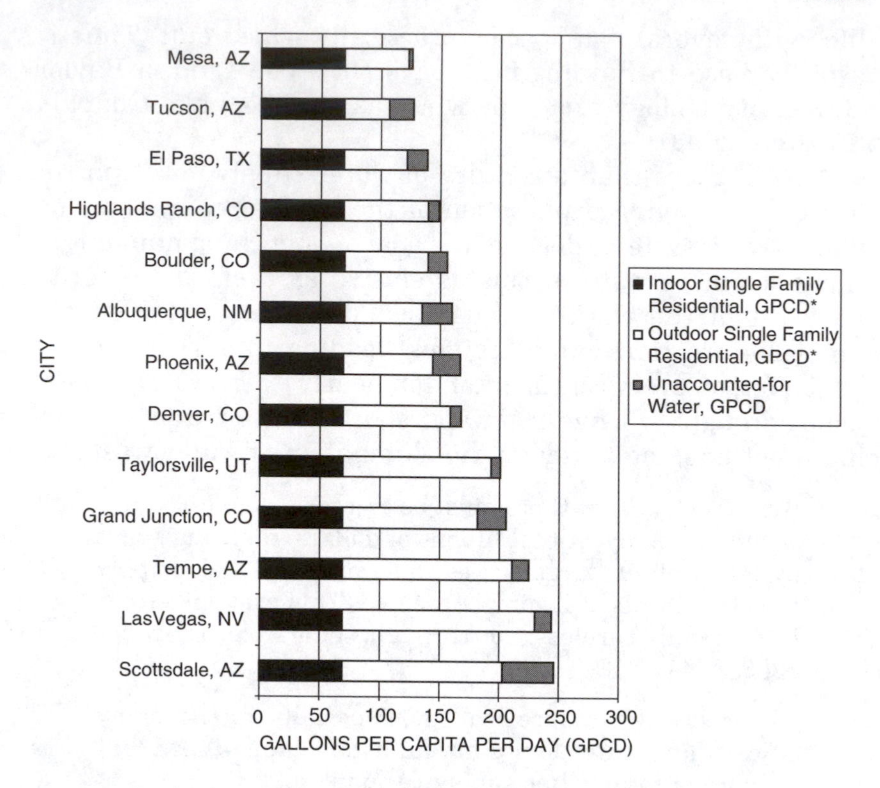

Figure 1 Per capita indicators of single-family water use and system unaccounted-for water in southwestern and western U.S. cities, 2001. (From Western Resource Advocates, 2003.)
* Estimated component of reported GPCD.

in Scottsdale, Arizona, or Las Vegas, Nevada, really *need* to use twice as much water as one in Mesa or Tucson, Arizona, with a virtually identical climate—and in a desert? (Figure 1).

Water waste and delayed drought management that resist calls for large-scale and aggressive conservation action hurt economies, too. Tourism, recreational, and related sales losses in Colorado in 2002, the same year Colorado experienced one of its worst droughts on record, were estimated at $1.7 billion, or 20% of normal, according to one study. Low water flows on the Colorado and Arkansas rivers in that state affected rafting and related recreational industries particularly hard (Cada, 2003), yet some cities and towns that draw from those and other water sources waited until the end of summer to impose their most stringent restrictions on non-essential, discretionary uses such as lawn watering. The establishment of *earlier* and *more aggressive conservation requirements,* particularly for landscape watering, could have better preserved streamflows and reservoir levels. For example, the reservoirs for Denver, Colorado, which draw partly from the Colorado River, were more than half empty before Denver Water mandated a "no watering" ban on October 1 (Gardener, 2004), just as the cooler days of autumn were arriving and outdoor watering was waning anyway. By then, the damage had been done. With its water levels still precipitously low, in late 2002 Denver Water began a $0.7 million cloud seeding program to increase its reservoir levels (*U.S. Water News*, 2003) in an attempt to help make up for what its water conservation program lacked. A recent study of single-family water use in Denver found that more than 55% is estimated to be used outdoors—primarily for lawn watering (Western Resource Advocates, 2003). The opportunity for significant water savings from this water use excess is obvious yet largely ignored.

While some point to the West and Southwest regions of the United States as examples of water mismanagement and misuse, unfortunately, such practices are becoming more prevalent, including in regions such as precipitation-rich New England. And they are taking a toll. Such demands can tax the ecological balance of reservoirs, rivers, and aquifers even during times of normal precipitation, but they incur even more severe impacts during drought. For example, the Ipswich River in eastern Massachusetts now runs dry periodically during the summer months because of excessive water withdrawals for suburban lawn irrigation that are diminishing that river's base flows. The Ipswich River actually dried up completely in 1995, 1997, and 1999 (Postel and Richter, 2003), leaving dead fish, ruined wildlife habitats, and a dry riverbed torn up by teenagers driving all-terrain vehicles. Although some argue that raising water rates and sending a strong pricing signal about the value of water will curb abusive water use, some people, particularly the affluent, are price insensitive when it comes to wanting a perfect-looking green lawn. As Postel and Richter (2003) point out in *Rivers for Life: Managing Water for People and Nature,* "hefty water bills may not be enough: outright bans on lawn watering when river flows drop below ecological thresholds may be necessary" (p. 176) to preserve healthy streamflows and fish stocks. Despite the reluctance of some public officials to curb excessive lawn watering, *Lawn Care for Dummies* expresses a core value of the water-wise: "Face it, you have more important things to do with water than put it on a lawn" (Walheim, 1998).

On the spectrum of water use, how wide and avoidable is the stretch of inefficiency and waste? When we compute the simple equation that subtracts our *true water needs* from our *total water demands,* the sum—water waste and inefficiency—reveals an expansive "new" source of freshwater capacity that can not only relieve the effects of drought but also help offset the adverse impacts of long-term shortages.

II. WATER CONSERVATION: THE GREAT UNTAPPED WATER SUPPLY

Water conservation is a powerful yet underutilized drought mitigation tool that can stave off the severe water shortages, financial losses, and public safety risks that historically have been assumed to be an inevitable consequence of drought. Hundreds of hardware technologies and behavior-driven measures are available to boost the efficiency of water use: when implemented and put into action, they can drive down short-term as well as long-term water demands (Vickers, 2001).

For nearly every example of water waste and inefficiency that can be found in water systems, homes, landscapes, industries, businesses, and farms, there is a water conservation device, technology, or practice that will save water (Table 1) (American Water Works Association, 1996; Postel, 1999; Smith and Vickers, 1999; Vickers, 2001). Hardware measures, such as leak repairs, low-volume toilets, and more efficient cooling and heating systems, will result in long-term demand reductions and typically require one action only (installation or repair) to realize ongoing water savings. Behavior-oriented measures, such as turning off the faucet while brushing teeth, and other actions involving human decision making, typically realize savings on a short-term basis but not over the long term. Because behavior-oriented conservation measures often

TABLE 1 Overview of Water Conservation Incentives, Measures, and Potential Savings

End User Category	Examples of Conservation Incentives & Measures	Potential Water Savings Range (%)[a]
System (water utility)	Low volume of system unaccounted-for water (maximum 10% of total production)	Varies
	System audit	
	Ongoing leak detection, repair, water loss control, and revenue recovery	
	Metering and meter maintenance (e.g., correct sizing, calibration, timely replacement)	
	Pressure regulation	
Residential (indoor)	Conservation-oriented rates, rebates, and program and policy incentives	10–50
	Toilets and urinals (low-volume, nonwater, composting, retrofit devices)	
	Showerheads and faucets (e.g., low-volume, aerators, retrofit devices)	
	Clothes washers and dishwashers (e.g., high-efficiency, full loads only)	
	Point-of-use hot water heaters (e.g., homes with high hot water losses)	
	Leak repair and maintenance (e.g., leaking toilets and dripping faucets)	
Lawn & landscape irrigation	Conservation-oriented rates, rebates, and program and policy incentives	15–100
	Water-efficient landscape design (e.g., functional turf areas only)	
	Native and/or drought-tolerant turf and plants (noninvasives only)	
	Limited or no watering of turf and landscape areas (beyond plant establishment)	
	Efficient irrigation systems and devices (e.g., automatic rain shut-off, drip hose for gardens)	
	Minimal or no fertilizers and chemicals (e.g., to control excessive growth and "watering in")	
	Rainwater harvesting (e.g., essential uses and efficient irrigation only)	
	Leak repair and maintenance (e.g., broken sprinkler heads and hoses)	

TABLE 1 Overview of Water Conservation Incentives, Measures, and Potential Savings (continued)

End User Category	Examples of Conservation Incentives & Measures	Potential Water Savings Range (%)[a]
Commercial, industrial, & institutional	Conservation-oriented rates, rebates, and program and policy incentives	15–50
	Submetering	
	Efficient cooling and heating systems (e.g., recirculating, point-of-use, green roofs)	
	Process and wastewater reuse, improved flow controls	
	Efficient fixtures, appliances, and equipment	
	Point-of-use hot water heaters (e.g., sites with large hot water losses)	
	Leak repair and maintenance (e.g., hose repair, broom and other dry cleaning methods)	
Agricultural	Conservation-oriented rates, rebates, and program and policy incentives	10–50
	Metering of on-farm water uses (e.g., irrigation, livestock)	
	Efficient irrigation systems and practices (e.g., surge valves, micro-irrigation, drip, LEPA, laser leveling, furrow diking, tailwater reuse, canal and conveyance system lining and management)	
	Efficient irrigation scheduling (e.g., customized, linked to soil moisture, local weather network)	
	Land conservation methods (e.g., conservation tillage, organic farming, Integrated Pest Management)	

[a] Actual water savings by individual users will vary depending on existing efficiencies of use, number and type of measures implemented, and related factors.

Sources: AWWA Leak Detection and Accountability Committee (1996), Postel (1999), Smith and Vickers (1999), Vickers (2001).

yield only temporary water savings, hardware and technology-based efficiency measures are favored by conservation managers, whose goal is permanent, long-term water reductions (Vickers, 2001). Case studies of efficiency measures implemented by individual end users among each major customer sector document not only water reductions, but also financial savings and other benefits (Table 2) (Adler et al., 2004; Bormann et al., 2001; DeOreo et al., 2004; Kenney, 2004; Ng, personal communication, 2003; U.K. Environment Agency, 2003).

The nearly 50% water demand reductions achieved by the city of Cheyenne, Wyoming, during record-breaking heat and minimal rain in the summer of 2002 exemplify how adherence to simple and reasonable conservation practices can enable a drought-stricken water supply system to stay robust. According to Clint Bassett, Cheyenne's water conservation specialist, "We encourage everyone to keep conserving water" (*WaterTech E-News*, 2003). Lawn watering restrictions during one month alone—July 2002—lowered average demand to 18.1 million gallons (68.5 megaliters) per day (mgd) compared to 34 mgd (128.7 megaliters) for the same month in the previous year—a 15.9 mgd (60.2 megaliters) savings. Further, Cheyenne's reservoirs were 83.5% full in the summer of 2002 compared to 63% the previous year without conservation. Cheyenne's conservation program results created a water reserve or bank that enabled it to better withstand even more severe drought conditions had they occurred.

The implementation of water efficiency options in response to drought and long-term water shortages demonstrates the profound role these strategies can serve in abating projected supply shortfalls. Beyond temporary drought responses, in some cases the water demand reductions from multi-year conservation programs have served to minimize or cancel major water and wastewater infrastructure expansion plans and related long-term capital debt. For example, the average 25% system-wide demand reductions realized by the Massachusetts Water Resources Authority (MWRA) in the early 1990s as a result of a comprehensive and multi-year conservation program have been maintained for more than a

Table 2 Examples of Water Savings from Conservation

End User Category	Measures Implemental	Reported Savings
System (water utility)	Water loss & leak reduction (Singapore): Reductions in unaccounted-for water (UFW) achieved through aggressive leak detection and repair, pipe renewal, and 100% metering (including the fire department). Active commercial, industrial, and residential meter replacement ensures accurate billing and minimization of unmetered water losses. Nonpotable water by industry is promoted and illegal connections can incur fines up to $50,000 or 3 years in prison.	System UFW reduced from 11% in 1989 to 5% by 2003, saving more than $26 million in avoided capital facility expansions
Residential (indoor)	Home building (Gusto Homes, England): Rainwater harvesting system and underground storage installed in 24 homes as well as dual-flush toilets, aerated showerheads, and solar water heaters.	Average 50 m^3/year per household water savings (50%)
Lawn & landscape irrigation	Native plants and natural landscaping (CIGNA Corporation, Bloomfield, CT): Conventional 120-ha corporate lawn converted to meadows, wildflower patches, and walking areas by the CIGNA Corporation (Bloomfield, CT) .	Several hundred thousand dollars savings per year in reduced water demands, fertilizer, pesticide, and equipment and maintenance needs; estimated conversion cost was $63,000
	Municipal drought lawn watering restrictions (8 municipal water providers in Colorado, U.S.): Outdoor watering restrictions were monitored to measure water savings achieved (comparison of 2002 drought year use to 2000/2001 average use), with the following results:	

	Once/week maximum mandatory lawn watering restriction (Lafayette, CO)	53% net water savings (average)
	Twice/week maximum mandatory lawn watering restriction (Boulder, CO; Fort Collins, CO; Louisville, CO)	30% net water savings (average)
	2⅓ times/week (once every 3 days) maximum mandatory lawn watering restriction (Aurora, CO; Denver Water, CO; Thornton, CO; Westminster, CO)	14% net water savings (average)
	Voluntary lawn watering schedules (Boulder, CO; Thornton, CO)	No water savings (average); net *increase* in water use
Commercial, industrial, & institutional	Supermarkets (6 supermarket sites in Southern California, U.S.): Advanced water treatment systems reduced fresh water needs for cooling systems. Other recommended efficiency measures included: high-efficiency spray nozzles, aerators, and flow restrictors installed on hand sinks and spray tables; elimination of garbage grinders, to be replaced by composting food wastes; and installation of high-pressure sprayers to replace low-pressure hoses for the meat department.	2,700 m³/year average water savings per supermarket
	Prison (Georgia Department of Corrections, Reidsville, Georgia, U.S.): Canning operation for vegetables (beans, carrots, greens, peas, potatoes, and squash) retrofitted with flowmeters, totalizers, and control valves to monitor water use. One rinse step eliminated and a counterflow rinsing system was installed to reduce freshwater requirements for cleaning vegetables. Alternative cooling system eliminated single-pass cooling water. Dry cleaning methods replaced water cleaning practices for floors and some equipment.	94,600 m³/year average water savings (about 57% of peak daily use); capital cost of measures was $38,000 and estimated savings are $102,700; simple payback less than 1 year

Table 2 Examples of Water Savings from Conservation (continued)

End User Category	Measures Implemented	Reported Savings
Agricultural	Dairy (United Milk Plc, England): Zero water use is the result of a reverse osmosis (RO) membrane system that was installed to recover and treat milk condensate for reuse throughout the plant.	657,000 m^3/year; $405,000 per year
	Produce (Unigro, Plc, England): Producer of pesticide-free fresh fruit, vegetables, and herbs uses precision irrigation and rainwater harvesting in a sealed, climate-controlled facility that requires 30% less water per unit of crop yield than conventional irrigation.	9,000 to 18,000 m^3/year (50%) average water savings; $7,400 per year

Sources: Adler et al. (2004), Bormann et al. (2001), DeOreo et al. (2004), Kenney (2004), Ng, personal communication (2003), U.K. Environment Agency (2003).

decade, and they are projected to continue. Instrumental to this achievement were aggressive leak repair (the city of Boston could not account for approximately 50% of its water during some of the 1980s), innovations in industrial water use efficiency, and the installation of water-saving toilets and plumbing fixture retrofit devices. These conservation savings not only transformed that system's supply status from shortfall to abundance, but they averted construction of a controversial dam project on the Connecticut River that was projected to incur a debt of more than $500 million (1987 dollars) to more than 2 million-plus residents and businesses in metropolitan Boston (Amy Vickers & Associates, Inc., 1996). Should the MWRA need to reduce demands even further (i.e., respond to a drought, supply new users, or meet emergency water demands), a plethora of additional water efficiency measures can be implemented to increase water savings beyond the 25% already realized.

Water use reductions from conservation can be especially significant when drought response combines with multi-year conservation programs. For example, during a drought in 2001, the city of Seattle, Washington, provided water use curtailment messages to the public (in addition to existing conservation measures) and had a significant impact on water demand in 2002, yielding 1.2 mgd (4.5 megaliters) in new long-term savings. These reductions surpassed the city's 2002 water savings goals by 8%. Seattle's continuing water conservation program ("1% program"), which has a 1% per year water reduction goal to lower demand 30% by 2010, has thus far realized a 20% decline in per capita use. Seattle's savings are considered long term because they include hardware-based, more permanent efficiency measures such as system leak reduction; financial incentives for industries, commercial establishments, and institutional users that install recirculated cooling and efficient-process water systems; rebates for the installation of low-volume (6 liters per flush) toilets; high-efficiency clothes washers; and discounts for natural yard care products that minimize lawn watering (Seattle Public Utilities, 2003).

Water conservation should not be just an emergency response to drought, but a long-term approach to managing and alleviating stresses on the world's finite water supplies. The significant water savings potential from large-scale conservation programs is increasingly recognized as an alternative to conventional (and costly) water supply development projects, including desalination and wastewater reclamation facilities. In a research study by the National Regulatory Research Institute, research specialist Melissa J. Stanford (2002) affirmed a similar view:

> Distribution system improvements, leak detection and remediation programs, water utility consolidation, wholesale purchasing agreements, demand management and integrated water resources planning, requests to conserve and water use restrictions, drought management planning and drought pricing, rate design alternatives, and communication and education are among the ways to bolster water supply and contend with drought. (p. 2)

In addition to the many benefits of conservation to drinking water systems, the recognition of ecological limits and the need to preserve streamflows through water efficiency and caps on use are also being incorporated into river and watershed schemes. For example, water extractions from the Murray-Darling river basin in Australia, that nation's largest and most economically important, have been capped to avert major damage to the river's ecological health. Even with the cap, the economy of that basin is projected to grow over the next 25 years (Postel and Richter, 2003), demonstrating that water efficiency is much more about boosting the productivity of water than sacrifice (Postel and Vickers, 2004).

Reducing water use is an obvious, in-kind response to drought and what nature presents: using less during times of shortfall, enjoying more in periods of natural abundance. "We all need to remember that water is not inexhaustible," remarks Bennett Raley (2004), assistant U.S. Department of Interior secretary for water and science. "Shortages will occur even in normal years. These shortages will threaten people, municipalities, farms, endangered species, and the environment. Doing nothing is not an option; it's not too early to start doing something about it now."

III. CONCLUSIONS

Water conservation is a powerfully effective short-term drought mitigation tool that is also an equally credible approach to better managing long-term water demands. Conservation-minded water systems have demonstrated that the efficient management of public, industrial, and agricultural water use during drought is critical to controlling and minimizing the adverse effects of reduced precipitation on water supplies. If we understand where and how much water is used and apply appropriate efficiency practices and measures to reduce water waste we can more easily endure—economically, environmentally, and politically—drought and projected water shortages. The lessons of effective drought management strategies—the implementation of comprehensive conservation measures—show that conservation can also be tapped to help overcome current and projected supply shortfalls that occur during non-drought times as well. The implementation of water waste reduction and efficiency measures can lessen the adverse impacts of excessive water demands on the natural water systems (rivers, aquifers, and lakes) and the ecological resources on which they depend. The notable demand reductions achieved by water efficiency–minded cities and water systems prove the significant role conservation can play in not only coping with drought but overcoming supply limitations and bolstering drought resistance through the preservation of water supply capacity. Like any savvy investor, efficiency-minded public officials and water managers who minimize their system water losses and invest in conservation will yield a treasure trove of "new" water to protect it from future shortages. Human activities play a key role in our experience of drought. A water-rich or water-poor future will be determined largely by our water waste and water efficiency actions *now.*

REFERENCES

Adler JA, K Mays, G Brown. Partnerships drive conservation in state government: A water efficiency success story for state prisons. Proceedings of the Water Sources Conference & Exposition, Austin, TX, January 11–14, 2004, American Water Works Association, Denver, CO, TUE8, pp. 1–4, 2004.

American Water Works Association. Leak Detection and Water Accountability Committee. Committee report: Water accountability. *Journal of the American Water Works Association* 88(7):108–111, 1996.

Amy Vickers & Associates, Inc. Final Report: Water Conservation Planning USA Case Studies Project. Prepared for the United Kingdom Environment Agency, Demand Management Centre. Amherst, MA, June 1996.

Annan K, U.N. Secretary-General. Towards a Sustainable Future. Presented at the American Museum of Natural History's Annual Environmental Lecture, New York, May 14, 2002.

Bormann FH, D Balmori, GT Gebelle. *Redesigning the American Lawn*, 2nd ed. New Haven, CT: Yale University Press, 2001.

Cada C. Colorado's tourism industry gets $10m government lift. *The Boston Sunday Globe* (pp. A10–A11). March 2, 2003.

Carlton J. A lake shrinks, threatening Mexican region. *The Wall Street Journal* (pp. B1, B8), September 3, 2003.

DeOreo WB, M Gentili, PW Mayer. Water conservation in supermarkets. Proceedings of the Water Sources Conference & Exposition, Austin, TX, January 11–14, 2004, American Water Works Association: Denver, CO, 2004.

Folwell AP. *Water-Supply Engineering*, 1st ed. New York: John Wiley & Sons, 1900.

Gardener L. Surviving the worst drought in 300 years. Presented at the Water Sources Conference & Exposition, Austin, Texas, January 11–14, 2004, pp 1–10.

Kenney DS. Colorado water restriction study. Proceedings of the Water Sources Conference & Exposition, Austin, TX, January 11–14, 2004, American Water Works Association, Denver, CO, 2004.

Ng HT, Public Utilities Board, Singapore, personal communication, September 2003.

Postel S. *Pillar of Sand: Can the Irrigation Miracle Last?* New York: W.W. Norton, 1999.

Postel S, B Richter. *Rivers for Life: Managing Water for People and Nature.* Washington, D.C.: Island Press, 2003.

Postel S, A Vickers. Boosting water productivity. In Worldwatch Instiitute: *State of the World 2004* (pp. 46–65). New York: W.W. Norton, 2004.

Raley B. Water 2025. Presented at AWWA's Water Sources Conference and Exposition, Austin, TX, January 12, 2004.

Ritter K. Western drought provides peek into flooded desert city. *The Boston Globe* (p. A20), August 31, 2003.

Seattle Public Utilities. *Regional 1% Water Conservation Program 2002 Annual Report.* Seattle, WA: Seattle Public Utilities, March 2003.

Smith JB, A Vickers. Unaccounted-for water: costs and benefits of water loss and revenue recovery in four Vermont municipal water systems. In Proceedings of CONSERV 99, Water Efficiency: Making Cents in the Next Century, Monterey, CA, January 31–February 3, 1999, American Water Works Association, Denver, CO, 1999.

Stanford MJ. Water supply assurance and drought mitigation: options for state regulatory commissions key stakeholders (executive summary), NRRI 02-13 es. National Regulatory Research Institute at The Ohio State University, Columbus, OH, November 2002.

United Kingdom Environment Agency. *2003 Water Efficiency Awards: Inspirational Case Studies Demonstrating Good Practice Across All Sectors.* Birmingham, England: The Environment Agency, (*www.environment-agency.gov.uk/savewater*, accessed August 2003), 2003.

United Nations. Population Division of the Department of Economic and Social Affairs of the United Nations Secretariat, World Population Prospects: The 2002 Revision and World Urbanization Prospects: The 2001 Revision, (*http://esa.un.org/unpp*, accessed January 2004), 2002.

United Nations. *Water for People, Water for Life: The United Nations World Water Development Report.* Paris: UNESCO Publishing and Berghahn Books, 2003.

U.S. Water News. In drought, Colorado seeds clouds and hopes snow is born. 20(1), 18, 2003.

Vickers A. *Handbook of Water Use and Conservation: Homes, Landscapes, Businesses, Industries, Farms.* Amherst, MA: WaterPlow Press, 2001.

Walheim L. *Lawn Care for Dummies*. New York: Hungry Minds, 1998.

WaterTech E-News. Water usage cut by 50 percent in Wyoming city. (*http://www.watertechonline.com/news.asp?mode=4&N_ID=42206*, accessed January 2004) August 11, 2003.

Western Resource Advocates. *Smart Water: A Comparative Study of Urban Water Use Efficiency across the Southwest.* Boulder, CO: Western Resource Advocates, December 2003.

8

The Role of Water Harvesting and Supplemental Irrigation in Coping with Water Scarcity and Drought in the Dry Areas

THEIB Y. OWEIS

CONTENTS

I. INTRODUCTION

Water scarcity and drought are among the most serious obstacles to agricultural development and a major threat to the environment in the dry areas. Agriculture in the dry areas accounts for more than 75% of the total consumption of water. With rapid increases in demand, water will be increasingly reallocated away from agriculture and the environment.

Despite scarcity, water continues to be misused. Mining groundwater is now a common practice, risking both water reserves and quality. Land degradation is another challenge in the dry areas, closely associated with drought-related water shortage. Climatic variation and change, mainly as a result of human activities, are leading to depletion of the vegetation cover and loss of biophysical and economic productivity. This happens through exposure of the soil surface to wind and water erosion and shifting sands, salinization of land, and water logging. Although these are global problems, they are especially severe in the dry areas.

Two major environments occupy the dry areas. The first is the wetter rain-fed areas, where rainfall is sufficient to support economical dry farming. However, because rainfall amounts and distribution are suboptimal, drought periods often occur during one or more stages of crop growth, causing very low crop yields. Variation in rainfall amounts and distribution from one year to the next causes substantial fluctuations in production. This situation creates instability and negative socioeconomic impacts. The second environment is the drier environment (steppe or *badia*), characterized by an annual rainfall too low to support economical dry farming. Most of the dry areas lie in this zone. Small and scattered rainstorms in these regions fall on lands that are generally degraded with poor vegetative cover. Rainfall, although low, may accumulate through runoff from vast areas in a large volume of ephemeral water and largely be lost through direct evaporation or in salt sinks.

With scarcity, it is essential that available water be used at highest efficiency. Many technologies are available to improve water productivity and management of scarce water resources. Among the most promising technologies are (1) supplemental irrigation (SI) for rain-fed areas and (2) rainwater harvesting (WH) for the drier environments (Oweis and Hachum, 2003). Improving scarce water productivity, however, requires exploiting not only water management but also other inputs and cultural practices. This chapter addresses the concepts and potential roles of supplemental irrigation and water harvesting in improving water productivity and coping with increased scarcity and drought in the dry areas.

II. SUPPLEMENTAL IRRIGATION

Precipitation in the rain-fed areas is low in amount and suboptimal in distribution, with great year-to-year fluctuation. In a Mediterranean climate, rainfall occurs mainly during the winter months. Crops must rely on stored soil moisture when they grow rapidly in the spring. In the wet months, stored water is ample, plants sown at the beginning of the season are in early growth stages, and the water extraction rate from the root zone is limited. Usually little or no moisture stress occurs during this period (Figure 1). However, during spring, plants grow faster, with a high evapotranspiration rate and rapid soil moisture depletion due to higher evaporative demand. Thus, a stage of increasing moisture stress starts in the spring and continues until the end of the season. As a result, rain-fed crop growth is poor and yield is low. The mean grain yield of rain-fed wheat in the dry areas is about 1 t/ha, far below the yield potential of wheat (more than 5–6 t/ha).

Supplemental irrigation aims to overcome the effects of drought periods as soil moisture drops and halts crop growth and development. Limited amounts of water, if applied during critical times, can result in substantial increases in yield and water productivity.

Research results from the International Center of Agricultural Research in the Dry Areas (ICARDA) and other organizations, as well as harvests from farmers' fields, have

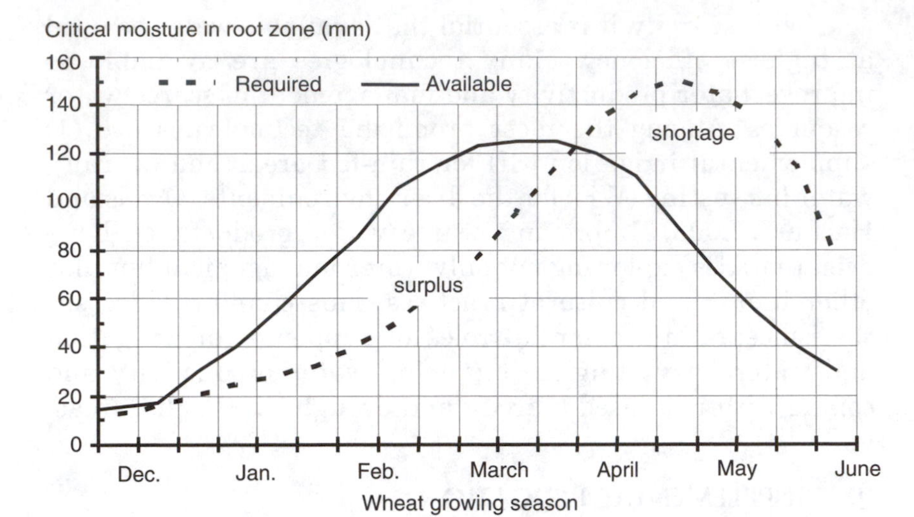

Figure 1 Typical soil moisture pattern over the growing season of a Mediterranean-type wheat. (From Oweis, 1997.)

demonstrated substantial increases in crop yield in response to the application of relatively small amounts of irrigation water. Table 1 shows increases in wheat grain yields under low, average, and high rainfall in northern Syria, with application of limited amounts of SI. By definition, rainfall is the major source of water for crop growth and production; thus the amount of water added by SI cannot by itself support economical crop production. In addition to yield increases, SI also stabilized wheat production over years (i.e., reduced the interannual variability of yields).

The impact of SI goes beyond yield increase to substantially improving water productivity. The productivity of irrigation water and rainwater is improved when they are used conjunctively (Oweis et al., 1998, 2000). Average rainwater productivity of wheat ranges from 0.35 to 1.0 kg/m^3. It was found that 1 m^3 of water applied as SI at the proper time could produce more than 2.0 kg of wheat.

Using irrigation water conjunctively with rain was found to produce more wheat per unit of water than if used alone in fully irrigated areas where rainfall is negligible. In fully irrigated areas, water productivity for wheat ranges from 0.5 to

TABLE 1 Yield and Water Productivity (WP) for Wheat under Rain fed and Supplemental Irrigation (SI) in Dry, Average, and Wet Seasons in Tel Hadya, North Syria

Season/Annual Rainfall (mm)	Rainfed Yield (t/ha)	Rainfall WP (kg/m^3)	Irrigation Amount (mm)	Total Yield (t/ha)	Yield Increase due to SI (t/ha)	Irrigation WP (kg/m^3)
Dry (234 mm)	0.74	0.32	212	3.38	3.10	1.46
Average (316 mm)	2.30	0.73	150	5.60	3.30	2.20
Wet (504 mm)	5.00	0.99	75	6.44	1.44	1.92

Source: Adapted from Oweis (1997).

about 0.75 kg/m^3, one-third of that achieved with SI. This difference suggests that allocation of limited water resources should be shifted to more efficient practices (Oweis, 1997). Food legumes, which are important for providing low-cost protein for people of low income and for improving soil fertility, have shown similar responses to SI in terms of yield and water productivity.

In the highlands of the temperate dry areas in the Northern Hemisphere, frost occurs between December and March. Field crops go into dormancy during this period. In most years, the first rainfall sufficient to germinate seeds comes late, resulting in a poor crop stand when the crop goes into dormancy. Rain-fed yields can be significantly increased if the crop achieves good early growth before dormancy. This can be achieved by early sowing with application of a small amount of SI. A 4-year trial, conducted at the central Anatolia plateau of Turkey, showed that applying 50 mm of SI to wheat sown early increased grain yield by more than 60%, adding more than 2 t/ha to the average rain-fed yield of 3.2 t/ha (ICARDA, 2003). Water productivity reached 5.25 kg grain/m^3 of consumed water, with an average of 4.4 kg/m^3. These are extraordinary values for water productivity with regard to the irrigation of wheat.

A. Optimization of Supplemental Irrigation

Optimal SI in rain-fed areas is based on the following three criteria: (1) water is applied to a rain-fed crop that would normally produce some yield without irrigation; (2) because rainfall is the principal source of water for rain-fed crops, SI is applied only when rainfall fails to provide essential moisture for improved and stable production; and (3) the amount and timing of SI are scheduled not to provide moisture stress–free conditions throughout the growing season, but to ensure a minimum amount of water available during the critical stages of crop growth that would permit optimal instead of maximum yield (Oweis, 1997).

1. Deficit Supplemental Irrigation

Deficit irrigation is a strategy for optimizing production. Crops are deliberately allowed to sustain some degree of water deficit and yield reduction (English and Raja, 1996). The adoption of deficit irrigation implies appropriate knowledge of crop water use and responses to water deficits, including the identification of critical crop growth periods, and of the economic impacts of yield reduction strategies. In a Mediterranean climate, rainwater productivity increased from 0.84 to 1.53 kg grain/m^3 of irrigation water when only one-third of the full crop water requirement was applied (Figure 2). It further increased to 2.14 kg/m^3 when two-thirds of the requirement was applied, compared to 1.06 kg/m^3 at full irrigation. The results show greater water productivity at deficit than at full irrigation. Water productivity is a suitable indicator of the performance of irrigation management under deficit irrigation of cereals (Zhang and Oweis, 1999), in analyzing the water saving in irrigation systems and management practices, and in comparing different irrigation systems.

There are several ways to manage deficit irrigation. The irrigator can reduce the irrigation depth, refilling only part of the root zone soil water capacity, or reduce the irrigation frequency by increasing the interval between successive irri-

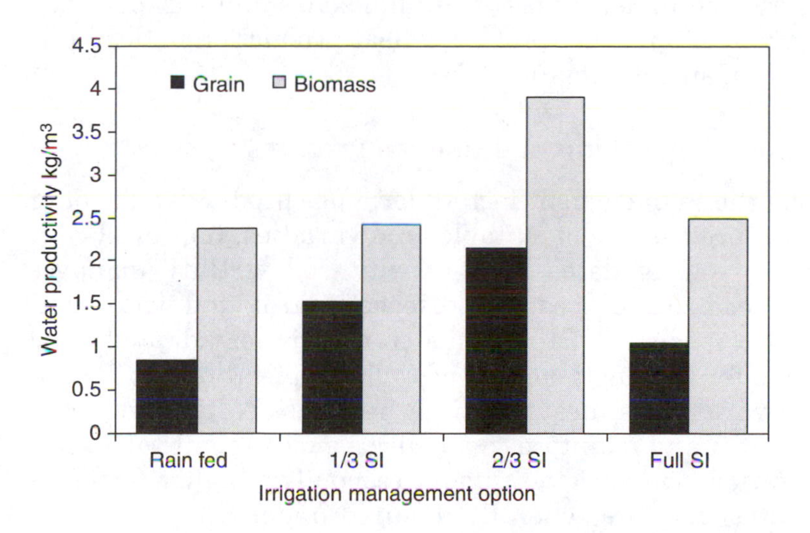

Figure 2 Water productivity of wheat under rain fed, deficit, and full SI conditions. (Adapted from Oweis, 1997.)

gations. In surface irrigation, wetting furrows alternately or placing them farther apart is one way to implement deficit irrigation. However, not all crops respond positively to deficit irrigation. This should be examined for local conditions and under different levels of water application and quality.

2. Maximizing Net Profits

An increase in crop production per unit of land or per unit of water does not necessarily increase farm profit because of the nonlinearity of crop yield with production inputs. Determining rain-fed and SI production functions is the basis for optimal economic analysis. SI production functions for wheat (Figure 4) may be developed for each rainfall zone by subtracting rainwater production function from total water production function. Because the rainfall amount cannot be controlled, the objective is to determine the optimal amount of SI that results in maximum net benefit to the farmers. Knowing the cost of irrigation water and the expected price per unit of the product, we can see that maximum profit occurs when the marginal product for water equals the price ratio of the water to the product. Figure 5 shows the amount of SI to be applied under different rainfall zones and various price ratios to maximize net profit of wheat production under SI in a Mediterranean climate.

3. Cropping Patterns and Cultural Practices

Among the management factors for more productive farming systems are the use of suitable crop varieties, improved crop rotation, sowing dates, crop density, soil fertility management, weed control, pest and disease control, and water conservation measures. SI requires crop varieties adapted to or suitable for varying amounts of water application. An appropriate variety manifests a strong response to limited water application and maintains some degree of drought tolerance. In addition, the varieties should respond to higher fertilization rates than are generally required under SI.

Given the inherent low fertility of many dry-area soils, judicious use of fertilizer is particularly important. In northern Syria, 50 kg N per hectare is sufficient under rainfed conditions. However, with water applied by SI, the crop responds to nitrogen up to 100 kg/ha, after which no further benefit is obtained. This rate of nitrogen uptake greatly improves water productivity. There must also be adequate available phosphorus in the soil so that response to nitrogen and applied irrigation is not constrained.

To obtain the optimum output of crop production per unit input of water, the mono-crop water productivity should be extended to a multi-crop water productivity. Water productivity of a multi-crop system is usually expressed in economic terms such as farm profit or revenue per unit of water used. Although economic considerations are important, they are not adequate as indicators of sustainability, environmental degradation, and natural resource conservation.

B. Water vs. Land Productivity

Land productivity (yield) and water productivity (WP) are indicators for assessing the performance of supplemental irrigation. Higher water productivity is linked with higher yields. This parallel increase in yields and water productivity, however, does not continue linearly. At some high level of yield, greater amounts of irrigation water are required to achieve additional incremental yield increase. Water productivity of wheat (Figure 3) starts to decline as yield per unit of land increases above certain levels.

It is clear that the amount of water required to achieve yield increases above 5 t/ha is much higher than that needed at lower yield levels. It would be more efficient to produce only 5 t/ha with lower water application than to achieve maximum yield with application of excessive amounts of water. The saved water would be used more efficiently if applied to new lands. This, of course, applies only when water, not land, is the limiting resource and without sufficient water to irrigate all the available land.

The association of high water productivity values with high yields has important implications for crop management in achieving efficient use of water resources in water-scarce areas (Oweis et al., 1998). Attaining higher yields with increased water productivity is economical only when the increased gains in crop yield are not offset by increased costs

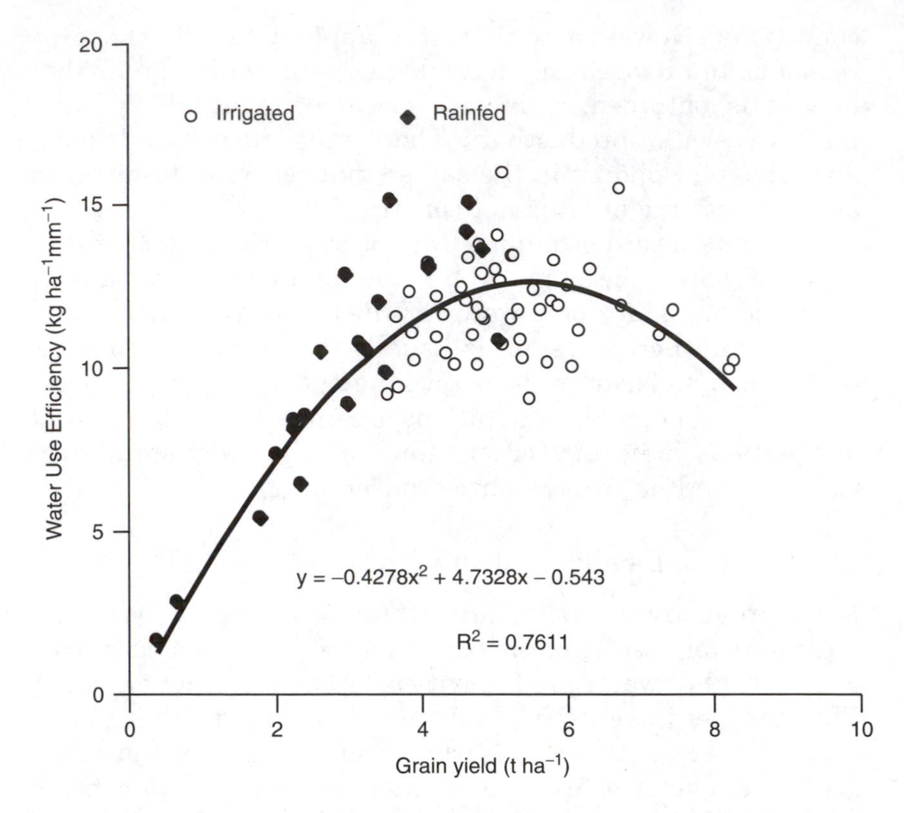

Figure 3 Relationship between crop water productivity and crop grain yield for durum wheat under SI in Syria. (From Zhang and Oweis, 1999.)

of other inputs. The curvilinear WP–yield relationship reflects the importance of attaining relatively high yields for efficient use of water. Policies for maximizing yield should be considered carefully before they are applied under water-scarce conditions. Guidelines for recommending irrigation schedules under normal water availability may need to be revised when applied in water-scarce areas.

III. WATER HARVESTING

A. The Concept and Components of the System

The drier environments, "the steppe," or, as they are called in the Arab world, *Al Badia*, occupy the vast majority of the dry

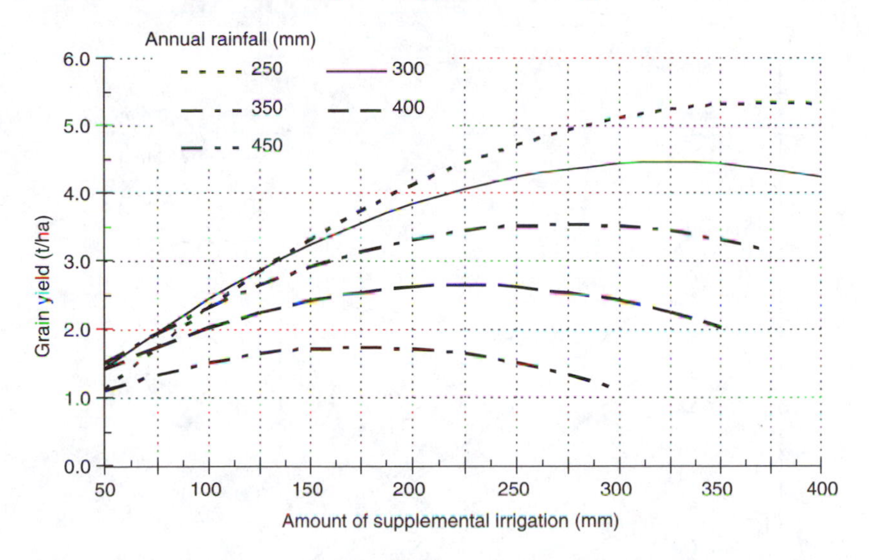

Figure 4 SI production functions for wheat in different rainfall zones in Syria. (Adapted from Oweis, 1997.)

areas. The disadvantaged people, who depend mainly on livestock grazing, generally live there. The natural resources of these areas are fragile and subject to degradation. Because of harsh natural conditions and the occurrence of drought, people increasingly migrate from these areas to the urban areas, with the associated high social and environmental costs.

Precipitation in the drier environments is generally low relative to crop requirements. It is unfavorably distributed over the crop-growing season and often comes with high intensity. It usually falls in sporadic, unpredictable storms and is mostly lost to evaporation and runoff, leaving frequent dry periods. Part of the rain returns to the atmosphere directly from the soil surface by evaporation after it falls, and part flows as surface runoff, usually joining streams and flowing to "salt sinks," where it loses quality and evaporates. A small portion of the rain joins groundwater. The overall result is that most of the rainwater in the drier environments is lost, with no benefits or productivity. As a result, rainfall in this environment cannot support economical dry farming like that in rain-fed areas (Oweis et al., 2001).

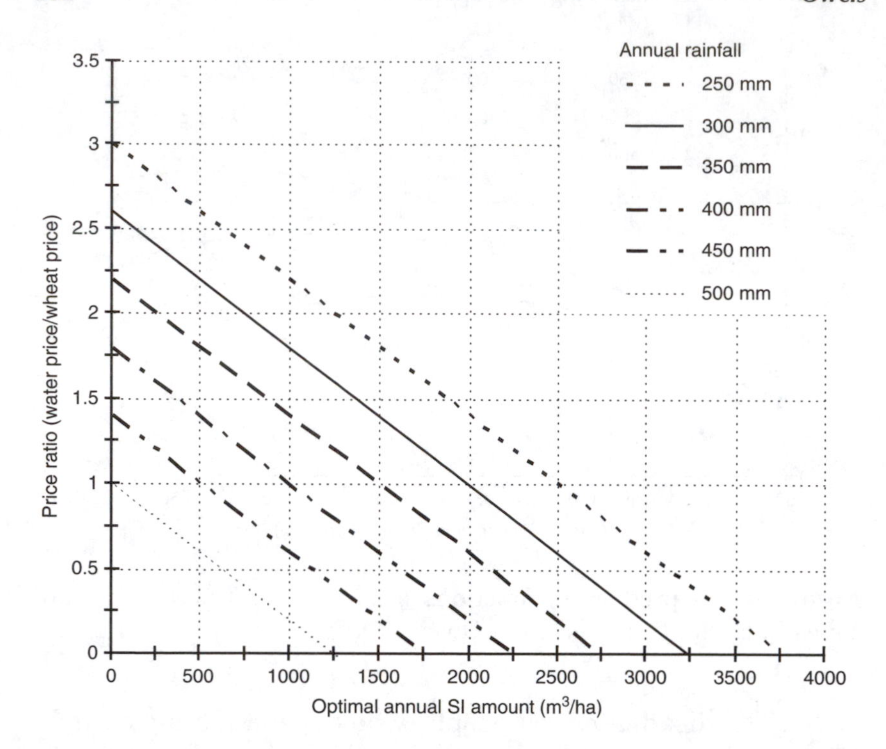

Figure 5 Optimal economical annual SI amount (m³/ha) in different rainfall zones in Syria. (Adapted from Oweis, 1997.)

Water harvesting can improve the situation and substantially increase the portion of beneficial rainfall. In agriculture, water harvesting is based on depriving part of the land of its share of rainwater to add to the share of another part. This brings the amount of water available to the target area closer to the crop water requirements so that economical agricultural production can be achieved. Water harvesting may be defined as "the process of concentrating precipitation through runoff and storing it for beneficial use."

Water harvesting is an ancient practice supported by a wealth of indigenous knowledge. Indigenous systems such as *jessour* and *meskat* in Tunisia; *tabia* in Libya; *cisterns* in north Egypt; *hafaer* in Jordan, Syria, and Sudan; and many other techniques are still in use (Oweis et al., 2004). Water harvesting may be developed to provide water for human and animal consumption, domestic and environmental purposes, and plant production. Water harvesting systems have three components:

1. *The catchment area* is the part of the land that contributes some or all of its share of rainwater to another area outside its boundaries. The catchment area can be as small as a few square meters or as large as several square kilometers. It can be agricultural, rocky, or marginal land, or even a rooftop or a paved road.
2. *The storage facility* is a place where runoff water is held from the time it is collected until it is used. Storage can be in surface reservoirs, in subsurface reservoirs such as cisterns, in the soil profile as soil moisture, or in groundwater aquifers.
3. *The target area* is where the harvested water is used. In agricultural production, the target is the plant or animal, whereas in domestic use, it is the human being or the enterprise and its needs.

B. Water Harvesting Techniques

Water harvesting techniques may be classified into two major types, based on the size of the catchment (Figure 6): micro-catchment systems and macro-catchment systems (Oweis et al., 2001).

1. Micro-Catchment Systems

Surface runoff in micro-catchment systems is collected from small catchments (usually less than 1000 m^2) and applied to an adjacent agricultural area, where it is stored in the root zone and used directly by plants. The target area may be planted with trees, bushes, or annual crops. The farmer has control, within the farm, over both the catchments and the target areas. All the components of the system are constructed inside the farm boundaries, which provides a maintenance and management advantage. But because of the loss of pro-

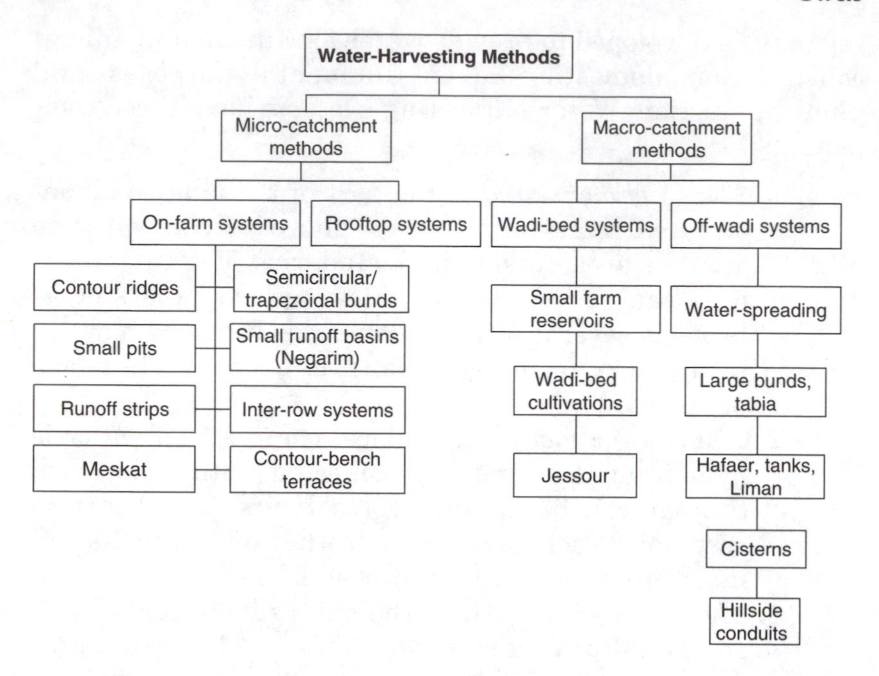

Figure 6 Classification of major rainwater harvesting systems in the dry areas. (From Oweis et al., 2001.)

ductive land it is practiced only in the drier environments, where cropping is so risky that farmers are willing to allocate part of their farm to be used as a catchment. They are simple in design and may be constructed at low cost. Therefore, they are easy to replicate and adapt. They have higher runoff efficiency than the macro-catchment systems and usually do not need a water conveyance system. Soil erosion may be controlled and sediment directed to settle in the cultivated area. These systems generally require continuous maintenance, with relatively high labor input. The most important micro-catchment water harvesting systems in the dry areas are described below.

a. *Contour Ridges*

Contour ridges consist of bunds, or ridges, constructed along the contour line at an interval of, usually, between 5 and 20 m. A 1- to 2-m strip upstream of the ridge is for cultivation, and the rest constitutes the catchment. The height of the ridges varies according to the slope and the expected depth of the runoff water retained behind it. The bunds may be reinforced by stones when necessary. This is a simple technique, which can be implemented by the farmers themselves. Bunds can be formed manually, with animal-driven equipment, or by tractors fitted with suitable implements. Ridges may be constructed on a wide range of slopes, from 1 to 50%.

Contour ridges are important for supporting the regeneration and new plantations of forage, grasses, and hardy trees on mild to steep slopes in the steppe (*badia*). In the semiarid tropics, they are used for the arable cropping of sorghum, millet, cowpeas, and beans. This system is sometimes combined with other techniques (such as the *zay* system) or with *in situ* water conservation techniques (such as the tied-ridge system) in the semiarid tropics.

b. Semicircular and Trapezoidal Bunds

Semicircular and trapezoidal bunds are earthen bunds created with spacing sufficient to provide the required runoff water for the plants. Usually, they are built in staggered rows. The technique can be used on an even, flat slope, but also on slopes up to 15%. The technique is used mainly for rangeland rehabilitation or fodder production, but can also be used for growing trees, shrubs, and, in some cases, field crops and vegetables.

c. Small Pits

The most famous pitting system is the *zay* system used in Burkina Faso. This form of pitting consists of digging holes 5–15 cm deep. Manure and grasses are mixed with some of the soil and put into the *zay*. The rest of the soil is used to form a small dike, down slope of the pit. Pits are used in combination with bunds to conserve runoff, which is slowed by the bunds. Pits are excellent for rehabilitating degraded agricultural lands. However, labor requirements for digging the *zay* are high and may constitute a large financial investment, year after year. This is because the pits have to be restored after each tillage operation. A special disk plow may be adjusted to create small pits for range rehabilitation.

d. Small Runoff Basins

Sometimes called *Negarim*, these runoff basins are small and of a rectangular or elongated diamond shape; they are surrounded by low earth bunds. *Negarim* work best on smooth ground, and their optimal dimensions are 5–10 m wide by 10–25 m long. They can be constructed on almost any slope, including very gentle ones (1–2% slopes), but on slopes above 5%, soil erosion may occur, and the bund height should be increased. They are most suitable for growing tree crops like pistachios, apricots, olives, almonds, and pomegranates, but they may be used for other crops. When used to grow trees, the soil should be deep enough to hold sufficient water for the entire dry season.

e. Runoff Strips

This technique is applied on gentle slopes and is used to support field crops in drier environments (such as barley in the *badia*), where production is usually risky or has a low yield. In this technique, the farm is divided into strips following contour lines. One strip is used as a catchment and the strip downstream is cropped. The cropped strip should not be too wide (1–3 m), and the catchment width should be determined with a view to providing the required runoff water to the cropped area. The same cropped strips are cultivated every year. Clearing and compaction may be implemented to improve runoff.

f. Contour Bench Terraces

Contour bench terraces are constructed on very steep sloping lands and combine soil-and-water conservation and water harvesting techniques. Cropping terraces are usually built to be level. Supported by stone walls, they slow water and control erosion. Steeper, noncropped areas between the terraces supply additional runoff water. The terraces contain drains to safely release excess water. They are used to grow trees and bushes but are rarely used for field crops. Some examples of this technique can be seen in the historic bench terraces in Yemen. Because they are constructed in steep mountain areas, most of the work is done by hand.

g. *Rooftop Systems*

Rooftop and courtyard systems collect and store rainwater from the surfaces of houses, large buildings, greenhouses, courtyards, and similar impermeable surfaces. Farmers usually avoid storing the runoff provided by the first rains to ensure cleaner water for drinking. If water is collected from soil surfaces, the runoff has to pass through a settling basin before it is stored.

The water collected is used mainly for drinking and other domestic purposes, especially in rural areas where there is no tap water. Extra water may be used to support domestic gardens. It provides a low-cost water supply for humans and animals in remote areas.

2. Macro-Catchment Systems

Macro-catchment systems collect runoff water from relatively large catchments, such as natural rangeland or a mountainous area, mostly outside farm boundaries, where individual farmers have little or no control. Water flows in temporary (ephemeral) streams called *wadi* and is stored in surface or subsurface reservoirs, but it can also be stored in the soil profile for direct use by crops. Sometimes water is stored in aquifers as a recharge system. Generally, runoff capture, per unit area of catchment, is much lower than for micro-catchments, ranging from a few percent to 50% of annual rainfall.

One of the most important problems associated with these systems involves water rights and the distribution of water, both between the catchment and cultivated areas and between various users in the upstream and downstream areas of the watershed. An integrated watershed development approach may overcome this problem. The most common macro-catchment systems are discussed below.

a. *Small Farm Reservoirs*

Farmers who have a *wadi* passing through their lands can build a small dam to store runoff water. The water can subsequently be used to irrigate crops or for domestic and animal consumption. These reservoirs are usually small, but may range in capacity from 1,000 to 500,000 m^3. The most important aspect of this system is the provision of a spillway with sufficient capacity to allow for the excessive peak flows. Most of the small farm reservoirs built by farmers in the rangelands (*badia*) have been washed away because they lacked spillway facilities or because their spillway capacity was insufficient. Small farm reservoirs are very effective in the *badia* environment. They can supply water to all crops, thus improving and stabilizing production. Moreover, the benefits to the environment are substantial.

b. Wadi-Bed Cultivation

Cultivation is very common in *wadi* beds with slight slopes. Because of slow water velocity, eroded sediment usually settles in the *wadi* bed and creates good agricultural lands. This may occur naturally or result from the construction of a small dam or dyke across the *wadi*. This technique is commonly used with fruit trees and other high-value crops. It can also be helpful for improving rangelands on marginal soils. The main problems associated with this type of water harvesting system are the costs and the maintenance of the walls.

c. Jessour

Jessour is an Arabic term given to a widespread indigenous system in southern Tunisia. Cross-wadi walls are made of either earth or stones, or both, and always have a spillway—usually made of stone. Over a period of years, while water is stopped behind these walls, sediment settles and accumulates, creating new land that is planted with figs and olives, but which may also be used for other crops. Usually, a series of *Jessour* are placed along the *wadi*, which originates from a mountainous catchment. These systems require maintenance to keep them in good repair. Because the importance of these systems for food production has declined recently, maintenance has also been reduced and many systems are losing their ability to function.

d. Water-Spreading Systems

The water-spreading technique is also called floodwater diversion. It entails forcing part of the *wadi* flow to leave its natural course and go to nearby areas, where it is applied to support crops. This water is stored solely in the root zone of the crops to supplement rainfall. The water is usually diverted by building a structure across a stream to raise the water level above the areas to be irrigated. Water can then be directed by a levee to spread to farms at one or both sides of the *wadi*.

e. *Large Bunds*

Also called *tabia*, the large bund system consists of large, semicircular, trapezoidal or open V-shaped earthen bunds with a length of 10 to 100 meters and a height of one to two meters. These structures are often aligned in long staggered rows facing up the slope. The distance between adjacent bunds on the contour is usually half the length of each bund. Large bunds are usually constructed using machinery. They support trees, shrubs, and annual crops but also support sorghum and millet in sub-Saharan Africa.

f. *Tanks and Hafaer*

Tanks and *hafaer* usually consist of earthen reservoirs, dug into the ground in gently sloping areas that receive runoff water either as a result of diversion from *wadi* or from a large catchment area. The so-called "Roman ponds" are indigenous tanks usually built with stonewalls. The capacity of these ponds ranges from a few thousand cubic meters in the case of the *hafaer* to tens of thousands of cubic meters in the case of tanks. Tanks are very common in India, where they support more than 3 million hectare of cultivated lands. *Hafaer* are mostly used to store water for human and animal consumption. They are common in West Asia and North Africa.

g. *Cisterns*

Cisterns are indigenous subsurface reservoirs with a capacity ranging from 10 to 500 m^3. They are basically used for human and animal water consumption. In many areas they are dug into the rock and have a small capacity. In northwest Egypt, farmers dig large cisterns (200–300 m^3) in earth deposits, underneath a layer of solid rock. The rock layer forms the ceiling of the cistern, whereas the walls are covered by impermeable plaster materials. Modern concrete cisterns are being constructed in areas where a rocky layer does not exist. In this system, runoff water is collected from an adjacent catchment or is channeled in from a more remote one. The first rainwater runoff of the season is usually diverted from the cistern to reduce pollution. Settling basins are sometimes constructed to reduce the amount of sediment. A bucket and rope are used to draw water from the cistern.

Cisterns remain the only source of drinking water for humans and animals in many dry areas, and the role they play in maintaining rural populations in these areas is vital. In addition to their more usual domestic purposes, cisterns are now also used to support domestic gardens. The problems associated with this system include the cost of construction, the cistern's limited capacity, and influx of sediment and pollutants from the catchment.

h. Hillside-Runoff Systems

In Pakistan, this technique is also called *sylaba* or *sailaba*. Runoff water flowing downhill is directed, before joining *wadi* by small conduits, to flat fields at the foot of the hill. Fields are leveled and surrounded by levees. A spillway is used to drain excess water from one field to another farther downstream. When all the fields in a series are filled, water is allowed to flow into the *wadi*. When several feeder canals are to be constructed, distribution basins are useful. This is an ideal system with which to utilize runoff from bare or sparsely vegetated hilly or mountainous areas.

C. Water Harvesting for Supplemental Irrigation

Where groundwater or surface water is not available for supplemental irrigation, water harvesting can be used to provide the required amounts during the rain season. The system includes surface or subsurface storage facilities ranging from an on-farm pond or tank to a small dam constructed across the flow of a *wadi* with an ephemeral stream. It is highly recommended when inter-seasonal rainfall distribution and/or variability are so high that crop water requirements cannot be reasonably met. In this case, the collected runoff is stored for later use as supplemental irrigation (Oweis et al., 1999). Important factors include storage capacity, location, and safety of storage structures. Two major problems associated with storing water for agriculture are evaporation and seepage losses. Following are management options proven to be feasible in this regard (Oweis and Taimeh, 2001):

1. Harvested water should be transferred from the reservoir to be stored in the soil as soon as possible after collection. Storing water in the soil profile for direct use by crops in the cooler season saves substantial evaporation losses that normally occur during the high evaporative demand period. Extending the use of the collected water to the hot season reduces its productivity because of higher evaporation and seepage losses.
2. Emptying the reservoir early in the winter provides more capacity for following runoff events. Large areas can be cultivated with reasonable risk.
3. Spillways with sufficient capacity are vital for small earth dams constructed across the stream.

IV. CONCLUSIONS

In the dry areas, where water is most scarce, land is fragile and drought can inflict severe hardship on already poor populations. Using water most efficiently can help alleviate the problems of water scarcity and drought. Among the numerous techniques for improving water use efficiency, the most effective are supplemental irrigation and water harvesting.

Supplemental irrigation has great potential for increasing water productivity in rain fed areas. Furthermore, it can be a basis for water management strategies to alleviate the effects of drought. Reallocating water resources to rainfed crops during drought can save crops and reduce negative economic consequences in rural areas. However, to maximize the benefits of SI, other inputs and cultural practices must also be optimized. Limitations to implementing supplemental irrigation include availability of irrigation water, cost of conveyance and application, and lack of simple means of water scheduling. In many places, high profits have encouraged farmers to deplete groundwater aquifers. Appropriate policies and institutions are needed for optimal use of this practice.

Water harvesting is one of the few options available for economic agricultural development and environmental protection in the drier environments. Furthermore, it effectively combats desertification and enhances the resilience of the communities and ecosystem under drought. Success stories are numerous and technical solutions are available for most situations. The fact that farmers have not widely adopted water harvesting has been attributed to socioeconomic and policy factors, but the main reason has been lack of community participation in developing and implementing improved technologies. Property and water rights are not favorable to development of water harvesting in most of the dry areas. New policies and institutions are required to overcome this problem. It is vital that concerned communities be involved in development from the planning to the implementation phases. Applying the integrated natural resource management approach helps integrate various aspects and avoid the conflicts of water harvesting and supplemental irrigation.

REFERENCES

English, M. and Raja, S.N., Perspectives on deficit irrigation, *Agri. Water Manage.*, 32, 1, 1996.

ICARDA, *ICARDA Annual Report 2002*, International Center for Agricultural Research in the Dry Areas (ICARDA), Aleppo, Syria, 2003.

Oweis, T., *Supplemental Irrigation: Highly Efficient Water-Use Practice*, ICARDA, Aleppo, Syria, 1997.

Oweis, T. and Hachum, A., Improving water productivity in the dry areas of West Asia and North Africa, in *Water Productivity in Agriculture: Limits and Opportunities for Improvement* (p. 179), Kijne, W.J., Barker, R., and Molden, D., Eds., CABI Publishing, Wallingford, U.K., 2003.

Oweis, T. and Taimeh, A., Farm water harvesting reservoirs: Issues of planning and management in the dry areas, in Proc. of a Joint UNU-CAS International Workshop, Beijing, China, 8–13 September 2001, United Nations University, 5-53-70 Jingumae, Sgibuya-ku, Tokyo-150-8925, pp. 165–183, 2001.

Oweis, T., Pala, M., and Ryan, J., Stabilizing rain-fed wheat yields with supplemental irrigation and nitrogen in a Mediterranean-type climate, *Agron. J.*, 90, 672, 1998.

Oweis T., Hachum, A., and Kijne, J., Water harvesting and supplemental irrigation for improved water use efficiency in the dry areas, SWIM Paper 7, International Water Management Institute, Colombo, Sri Lanka, 1999.

Oweis, T., Zhang, H., and Pala, M., Water use efficiency of rainfed and irrigated bread wheat in a Mediterranean environment, *Agron. J.*, 92, 231, 2000.

Oweis, T., Prinz, D., and Hachum, A., *Water Harvesting: Indigenous Knowledge for the Future of the Drier Environments*, ICARDA, Aleppo, Syria, 2001.

Oweis, T., Hachum, A., and Bruggeman, A., Eds., *Indigenous Water Harvesting Systems in West Asia and North Africa*, ICARDA, Aleppo, Syria, 2004.

Zhang, H. and Oweis, T., Water-yield relations and optimal irrigation scheduling of wheat in the Mediterranean region, *Agri. Water Manage.*, 38, 195, 1999.

9

Drought, Climate Change, and Vulnerability: The Role of Science and Technology in a Multi-Scale, Multi-Stressor World

COLIN POLSKY AND DAVID W. CASH

CONTENTS

I. INTRODUCTION

One of the principal concerns about anthropogenic climate change is a possible alteration of the hydrologic cycle, including an increased frequency or magnitude of droughts. The underlying assumption is that a diminished water supply will cause social, economic, and ecological hardships. This assumption makes sense for many places in the world, such as southern Africa, where water supply plays a crucial and obvious role in determining social welfare. In these places, drought is legitimately considered a top national planning priority. Yet for other places, such as the United States, droughts' contribution to large-scale hardships is less obvious relative to other environmental hazards. For example, in the arid city of Phoenix, Arizona, the 2002 drought was expected to require only a modest 5% reduction in water consumption—hardly a memorable society-wide impact (*The Economist*, 2003). In the United States, droughts have been termed (Wilhite, 2001), in reference to the famous U.S. comedian, the "Rodney Dangerfield" of environmental hazards: they get no respect! As a result, in the United States at least, drought planning largely favors response over preparedness measures, and plans are relatively uncoordinated (National Drought Policy Commission, 2000).

This state of affairs is puzzling given that droughts exact a larger financial toll than any other natural hazard nationwide (Wilhite and Wood, 2001). A partial explanation may be found in the success of recent social adaptations to the most pernicious effects of droughts, such as famine and economic collapse. For the better part of the 20th century, the United States has implemented a sustained research and development program designed to apply science and technology (S&T) to drought-sensitive sectors of society. As a result, the dramatic negative impacts experienced during the "Dust Bowl" (the 1930s) have not been repeated in subsequent droughts. No one disputes the progress made in terms of lives lost and other severe impacts from drought. Yet the other damages associated with drought continue to mount, as population grows and farming continues to expand in drought-prone areas.

This qualified success story of social adaptations to the effects of drought, in the United States as elsewhere, suggests the following research question:

> Do past successes of adaptation to drought suggest that S&T will rise to future challenges posed by a greenhouse-induced increase in drought severity, not only in places where drought is a central planning focus, but also in places where drought receives less coordinated planning attention?

The objective of this chapter is to outline an answer to this question by linking two emerging literatures: the analysis of social vulnerability to the effects of global change (e.g., Kelly and Adger, 2000; Polsky et al., 2003; Turner et al., 2003 and Schröter et al. 2005) and the role of science and technology institutions in natural resource management (e.g., Cash et al., 2003; Kates et al., 2001; World Bank, 1999). In Sections II and III, we introduce, respectively, the concept of global change vulnerability and associated assessments, and the notion of institutions as cause and consequence of (and solution to) the effects of global change. In Section IV, the general conceptual discussions are linked to the specific case of drought, using past research on the U.S. Great Plains as a motivating example. In Section V, we outline four characteristics of institutions that improve the chances of successful reductions in vulnerability. We conclude with suggestions on future research directions.

II. GLOBAL CHANGE VULNERABILITY AND VULNERABILITY ASSESSMENTS

The objective of global change vulnerability assessments is to prepare specific communities of stakeholders to respond to the effects of global change (Schröter et al., 2005). There is a growing call to favor "vulnerability" assessments over the more familiar "impacts" approach to research on the human dimensions of global environmental change (e.g., Downing,

2000; Kelly and Adger, 2000; Liverman, 2001; McCarthy et al., 2001; National Research Council, 1999; Parry, 2001; Turner et al., 2003). In this literature, vulnerability is generally defined as a function of exposure to stresses, associated sensitivities, and relevant adaptive capacities. Thus to be vulnerable to the effects of stresses associated with global change, human–environment systems must be not only exposed and sensitive to the changes but also unable to cope. Conversely, systems are (relatively) sustainable if they possess strong adaptive capacity (Finan et al., 2002). In the former case, some form of anticipatory action would be justifiable to mitigate the ecological, social, and economic damages anticipated from global change, whereas in the latter case there would be less reason for concern and action. Vulnerability assessments are therefore a necessary part of sustainability science, or basic research intended to protect social and ecological resources for present and future generations (Clark and Dickson, 2003; Kates et al., 2001).

The common distinction between the vulnerability and impacts perspectives is that the former emphasizes the factors that constrain or enable coupled human–environment systems to adapt to stress, whereas the latter focuses more on system sensitivities and stops short of specifying whether a given combination of stress and sensitivity will result in an effective adaptation. In fact, this distinction applies more to the empirical studies of climate change impacts than to the conceptual underpinnings. Adaptation has been at the heart of the debate on reducing vulnerability to environmental stresses for a long time (Turner et al., 2003). Even the early models from the climate change impacts canon (e.g., Kates, 1985) do not exclude the process of adaptation, and the same applies to the broader, related literatures on risk and hazards (e.g., Burton et al., 1978; Cutter, 1996; Kasperson et al., 1988) and food security (e.g., Böhle et al., 1994; Downing, 1991). Thus the recent explosion of interest in "global change vulnerability" is not so much the result of a revolution in ideas—although theories are developing (e.g., Adger and Kelly, 1999)—but a response to a general dissatisfaction with the ways in which adaptive capacity has been captured in empirical research and the associated need to reconnect with this concept if global change models are to improve.

Polsky et al. (2003) suggest that successful empirical research on global change vulnerability should satisfy the following five (overlapping) criteria: (1) exhibit a place-based focus; (2) devote equal energy to exploring future trends and historical events; (3) treat stresses as multiple and interacting instead of unique or multiple and independent; (4) include not only natural and social science but also local ("indigenous" or "user-specific") knowledge; and (5) examine how adaptive capacity varies both within and between populations. This last criterion is especially important for defining vulnerability in the case of drought. In the United States at least, institutions that regulate on the one hand and design and disseminate new technologies on the other hand are the principal pathways for drought response, in anticipatory and reactive modes. To be sure, individual people do actively participate in drought mitigation activities, but the most important current set of options for adaptations to the effects of droughts, we argue, is associated with institutions (detailed in Sections III and IV).

It is difficult to specify quantitative models of how institutions enhance or reduce adaptive capacity. This difficulty is important in the climate change context because quantitative models, for better or worse, have occupied center stage in the debate on possible impacts from climate change and associated policy responses. The majority of these models are grounded in neoclassical economic theory, where the role of institutions in mediating impacts is largely if not entirely discounted. In these cases an individualistic perspective presumes that all people (modeled agents) are "economically rational." These modeled agents will implement any and all necessary adaptations to the effects of climate change "perfectly" (i.e., instantaneously and at greatest individual profit). In this way the role of institutions in influencing social response is implicitly assumed to be trivial, or, if significant, then of equal importance everywhere and always and as such not worthy of specifying in a model.

The canonical example of this approach is Mendelsohn et al.'s (1994) influential Ricardian analysis of climate change impacts in U.S. agriculture. This approach uses a regression model to evaluate the importance of climate in the determination of agricultural land values (in the contiguous United States) relative to other important factors such as population density and soil quality. The possible economic impacts of climate change are projected by multiplying the statistical relationships between historical climate and land values by a hypothetical climate change. Not surprisingly, the projected economic impacts based on the "perfect" adaptive capacity described above, defined in strict profit terms, are lower than in studies that do not allow the modeled agents to respond at all (i.e., where adaptive capacity is assumed a priori to be null).

Of course, if it is unrealistic to assume that agents possess no adaptive capacity, then it is equally unrealistic to assume that they possess perfect adaptive capacity. For example, the decisive factor behind a farmer's choice to prepare for drought through summer fallowing or portfolio diversification may hinge on the advice of an agricultural extension agent—who may or may not have the farmer's profit maximization as the number one priority (Riebsame, 1983). Thus, in principle, greater realism can be achieved by incorporating in the models some of the missing institutional landscape (Hanemann, 2000). Institutional influences should be particularly important in regions where climate change results in a strengthened drought regime.

Polsky (2004) modified the basic Ricardian framework to explore how institutions modulate agricultural climate sensitivities. In this analysis of agricultural land values in the U.S. Great Plains, statistical relationships are estimated at multiple spatial scales simultaneously: for the region as a whole ($n = 446$ counties); for the meso-scale (two subregions defined by the boundaries of the Ogallala Aquifer: $n_1 = 209$, $n_2 = 237$); and for the micro-scale (many sets of small numbers of counties; $n \approx 7$ on average). For each of the 6 years analyzed, the regression model fit better for the subregion defined by the boundaries of the Ogallala Aquifer than for the rest of the Great Plains. These differences in model fit were modest in

1969, dramatic in 1974, 1978, and 1982, and intermediate in 1987 and 1992, and they suggest that unspecified factors are responsible for buffering fluctuations in land values in the Ogallala relative to the rest of the Great Plains. The Ogallala is characterized by strong S&T and natural resource management institutions developed in response to the challenge of drought and the opportunity of irrigation. Thus an emerging hypothesis is that differences in the form and function of these institutions between the two subregions explain differences in the climate sensitivities of the two subregions (see also Emel and Roberts, 1988). Clearly, testing this hypothesis requires an in-depth study of the ways in which such institutions produce and disseminate knowledge.

III. INSTITUTIONS AND GLOBAL–REGIONAL–LOCAL ENVIRONMENTAL CHANGE

As discussed in the introduction, many adaptations have been implemented in U.S. water management in recent decades, but the development of institutions that conduct research, assessment, and technology development may be among the most influential. These developments have been neither unqualified successes nor unmitigated disasters. Instead, the results have been mixed. Thus what we need to identify and reduce social vulnerability to the effects of drought is a systematic understanding of which institutional designs lead to effective water management in the face of stress, whether in the form of anticipatory mitigation actions, post hoc reactions, or both (Cash et al., 2003).

A rich literature exists on the role of institutions in modulating human behavior in general. And there is a smaller but growing literature on the specific topics of how institutions link (a) science and technology to natural resource and environmental management and (b) actors across levels of organization. From markets, to international treaties, to norms and procedures of peer review, institutions have been instrumental in helping societies organize collective and individual action. As formal and informal systems of rules and decision-making procedures that guide social practices, and institutions have played important roles in international environmental governance, national efforts to address issues such as drought, and local management efforts (Keohane and Levy, 1996; Keohane et al., 1994; Young, 1999). But how do institutions relate to S&T production and use, and how, for issues such as drought management, do institutions influence multilevel dynamics?

Knowledge in general, and the productions and use of S&T information in particular, have become increasingly important forces shaping the course of international through local affairs (Clark et al., in review; Keohane and Nye, 1989; Sachs, 2001; World Bank, 1999). Technical information, in the form of factual knowledge about the state of the world and causal theories about how it works, is increasingly called on to guide tasks ranging from verifying nuclear testing treaties, to planning structural adjustment policies for struggling economies, to managing underground aquifers. A belief in the potential power of information has led to calls for improved transparency of information flows (Mitchell and Bernauer, 1998).

But the vague recognition that information matters has not led to agreement on when, how, and under what conditions it influences the behavior of policy actors. Despite the vast and growing array of institutions involved in collecting, analyzing, and disseminating information potentially relevant to global through local governance, our understanding of the role that these "information institutions" play remains limited (Keohane and Nye, 1989; Nye and Donahue, 2000). Despite this limitation, some notions are emerging. The influence of information depends on the form of institutions, their degree of formalization, and the pathways by which they process information. Some influence the production of scientific knowledge directly through norms and procedures regarding setting research priorities, targeting resources, conducting experiments, ensuring quality control (e.g., through peer review), and disseminating results. Others guide the preparation and dissemination of scientific information to a range of audiences, from the international consortium of weather services to international environmental data collection collaborations. Other institutions create the norms and procedures of science advising, technology assessment, and formal scientific assessments. They produce public information for an audience that includes managers and decision makers engaged in behaviors and in promulgating policies directly involved in transboundary environmental issues.

In a recently concluded 5-year research effort, the Global Environmental Assessment Project focused on this third type of information institutions, drawing conclusions from more than 40 case studies on assessment efforts addressing climate change, biodiversity loss, ozone depletion, water management, and transboundary air pollution (Clark et al.). The research suggests four basic ideas:

1. Institutions that support scientific assessments can influence policy, but influential assessments are the exception rather than the rule. Even influential assessments rarely affect policy choice directly, but rather exert substantial indirect influence on long-term issue development, such as who participates, what policy goals are emphasized, and what gets public attention.
2. The most influential assessments are those that are simultaneously perceived by a broad array of actors to possess saliency, credibility, and legitimacy.
3. Institutions shape the influence of assessments in large part by shaping the tradeoffs among saliency, credibility, and legitimacy and providing the context within which those tradeoffs can be balanced by assessment designers.
4. Effective information institutions play boundary-crossing functions, consciously connecting science and policy arenas.

Points 2–4 are addressed in greater detail in Section V.

In addition to this literature on institutions, a more developed suite of literatures germane to this chapter is that of multi-level dynamics in management. Well-developed theories on the challenge of governance in a multi-level world have emerged in such fields as hierarchy theory (Levin, 1997; O'Neill, 1988; Simon, 1962), human geography (Easterling and Polsky, forthcoming), adaptive management (Gunderson et al., 1995; Holling, 1978), and environmental federalism (Esty, 1996; Kincaid, 1996). The research in three other fields is particularly important for our purposes.

First, an explicit component of institutional analysis in the common pool resources (CPR) literature identifies that what happens at one scale can provide institutional constraints or opportunities at other scales: "[A]ppropriation, provision, monitoring, enforcement, conflict resolution, and governance activities are organized in multiple layers of nested enterprises. ... Establishing rules at one level, without rules at the other levels, will produce an incomplete system that may not endure over the long run" (Ostrom, 1990, p. 101–102).

While addressing the multi-level nature of commons problems, however, this line of research often casts the problem in terms of a simple dichotomous decision choice between centralized (higher level) control and autonomous or local control (Adams, 1990; Avalos and DeYoung, 1995; Bruggink, 1992; Somma, 1994). Another vein of research in the CPR literature has provided more nuanced interpretations and has better conceptualized and analyzed scale (Blomquist, 1992; Ostrom, 1998). This line of research has begun to identify the importance of polycentric networks—distributed systems of governance with coordinated governing authorities that link actors and institutions at different levels and apportion roles to different nodes in the network, balancing the tradeoffs between centralized and autonomous decision making (McGinnis, 1999).

Second, recent work in the field of international relations has focused on the interactions of regimes at international, national, and subnational levels, with a special focus on what constraints and opportunities are imposed by institutions at one level on institutions at other levels (Keohane and Ostrom, 1995; Young, 1995):

> ... [I]t seldom makes sense to focus exclusively on finding the right level or scale at which to address specific prob-

> lems arising from human/environment relations. Although small-scale or local arrangements have well-known problems of their own, there are good reasons to be wary of the pitfalls associated with the view that the formation of regimes at higher levels of social organization offers straightforward means of regulating human activities. ... In most cases, the key to success lies in allocating specific tasks to the appropriate level of social organization and then taking steps to ensure that cross-scale interactions produce complementary rather than conflicting actions. (Young, 2002, p. 266)

Finally, with its foundations in adaptive management, resilience theory (a literature mirroring much of the vulnerability literature, but with few researchers in common) has begun to address the importance of institutions, especially in facilitating learning and adaptation. For problems that cut across levels, institutions that are decentralized but link higher level and lower level management actions are seen as more resilient to external and internal shocks (Berkes, 2002; Folke et al., 2002). According to this perspective,

> [D]ynamic efficiency is frequently thwarted by creating centralized institutions and enhanced by systems of governance that exist at multiple levels with some degree of autonomy complemented by modest overlaps in authority and capability. A diversified decision-making structure allows for testing of rules at different scales and contributes to the creation of an institutional dynamics important in adaptive management. (Folke et al., 2002, p. 21)

IV. DROUGHT, CLIMATE CHANGE, AND AGRICULTURE IN THE U.S. GREAT PLAINS

Despite the richness of this literature on institutions and environmental affairs, there is room for additional research on the institutional design criteria that lead to effective water management in general and drought management (whether anticipatory or reactive) in particular. We return to the case of the U.S. Great Plains. Substantial evidence exists about specific water management successes and failures in this region (e.g., Glantz, 1994; Riebsame, 1990; Riebsame, 1991; Webb, 1931). The fact that the region has experienced multiple droughts subsequent to the 1930s Dust Bowl years without the associated dramatic impacts on human health, soil quality, employment, and out-migration is generally taken as a reflection of the success of the social adaptations implemented in response to the event (Warrick and Bowden, 1981). Although it is important to recognize this general success, subsequent Great Plains droughts have reminded residents that sensitivity to the effects of droughts is a dynamic process, and that past successes are no reason to cease improving disaster preparedness (Popper and Popper, 1987; Rosenberg and Wilhite, 1983; Wilhite and Easterling, 1987) or risk management (Wilhite, 2001).

Adaptations in the United States since the 1930s have centered on government assistance, along two dimensions: insurance against losses from natural disasters, and science and technology outreach. Spectacular evidence of successful S&T outreach is seen in the spread of irrigation among Great Plains farmers. From the end of the 1940s through the end of the 1990s, irrigated acreage in the region expanded from 2.1 million acres to 13.9 million acres (McGuire, 2003). This increase was catalyzed by massive S&T outreach programs that promoted irrigation as a fix for the factor (rainfall) limiting agriculture in this semiarid region. The irrigation has been largely restricted to farmers with access to one of the continent's largest aquifers, the Ogallala Aquifer. The existence of the irrigation water has allowed for a large-scale, intensive agricultural system that drives local economies to flourish at levels it might not otherwise reach (Kromm and White, 1990). This fossil resource may be approaching the end of its economic life: in some parts of the region, withdrawal rates have exceeded recharge rates by a factor of 100 (Taylor et al., 1988). Irrigation efficiencies have been improving in recent years, but it remains to be seen if the resource is being used sustainably (Riebsame, 1991; Wilhite, 1988).

As unsustainable as this water mining may be, there is little reason to expect significant changes in irrigation rates. Since the 1930s, a "moral geography" has emerged at the national level vis-à-vis the Great Plains. Federal financial assistance has been repeatedly offered during times of stress to "needy Jeffersonian yeomen farmers" almost regardless of cost (Opie, 1998) and sometimes in spite of actual need (Wilhite, 1983). This regional social contract generally favors the growth-driven industrial model of agriculture over economic diversification or ecologically sensitive land uses (Riebsame, 1994; Roberts and Emel, 1995). Given these institutional biases and incentives and associated market forces, farmers with access to Ogallala water are almost forced to irrigate. As such, one of the most effective mechanisms for reducing sensitivity to drought—Ogallala irrigation—may cease to be viable at some point during the 21st century. Some areas of the Ogallala have even instituted a "planned depletion" water policy (White, 1994), thereby only postponing—not preventing—when substantial changes in the regional economy (e.g., abandoning intensive farming altogether) may have to be made. In conclusion, the remarkable reductions in vulnerability attributable to the adaptations undertaken by Great Plains farmers (and policy makers in Washington, D.C.) during the 20th century may prove to be only a short-term fix (Bowden et al., 1981; Hulett, 1981; Opie, 2000; Riebsame, 1991; Wilhite, 2001). Only time will tell if the combination of a declining Ogallala water table with a significant drought will overwhelm local institutions' ability to cope (Glantz and Ausubel, 1988; Wilhite, 1988).

V. DESIGNING INSTITUTIONS TO LEVERAGE SCIENCE AND TECHNOLOGY TO ACHIEVE SUSTAINABLE DEVELOPMENT

The research reviewed above provides a springboard for beginning to develop an understanding of the design characteristics of effective drought management institutions. Our research on climate sensitivities and water management in the Great Plains complements this work and suggests several propositions that might tie together a number of the literatures described in Section III.

A first proposition is that multiple boundaries characterize the landscape of drought assessment, planning, and management, and that a key role institutions can play in reducing vulnerability is to better manage such boundaries. Perhaps the most fundamental of these boundaries is that between science and policy, in which actors on both sides of the boundary have an interest in maintaining the separation of the two arenas (Gieryn, 1995): "To shore up their claims on cognitive authority, scientists have to impose their own boundaries between science and policy" (Jasanoff, 1987, p. 199). Scientists have an interest in maintaining a boundary to ensure the credibility of their work. Politicians have an interest in maintaining a boundary to ensure their claims of representative legitimacy. But although there is interest in maintaining this boundary, scientists also have an interest in bridging boundaries when they seek to have science contribute socially relevant information that can be used by policy makers and decision makers. Thus the trick is *managing* the boundary, bridging where necessary, but maintaining it as a barrier as well.

A similar tension exists for other important boundaries. Academic disciplines (and boundaries between them) exist to deepen understanding of issues using specific and agreed-upon tools, yet interdisciplinary research is needed to understand and solve problems characterized by interconnectedness. Boundaries exist across levels to balance the efficiencies of centralized governance and the specificity to local context, yet coordination across jurisdictions, from global to local levels, seems to be necessary to address transboundary problems, commons problems, and the interactions between global and local change (Cash and Moser, 2000; Wilbanks and Kates, 1999). Finally, boundaries also exist between different issue areas—many management regimes are structured by issue (e.g., water, agricultural, and environmental agencies in states), again, allowing efficient focus on a narrow topic, but providing barriers for integrated management.

Given this prevalence of boundaries in human systems, what does recent research tell us about institutional structures that facilitate the management of boundaries? An emerging literature grounded in the social studies of science characterizes such institutions as "boundary organizations": institutions that act as an intermediary across boundaries and provide functions of convening, translation, collaboration, and mediation (Cash, 2001; Guston, 1999; Guston, 2001). Such organizations act as the site of co-production of knowledge, where scientists, managers, decision makers, and users of information jointly set agendas and decide on appropriate methodologies and products (Andrews, 2002).

In the Great Plains, for water and other natural resource issues, boundary organizations are embodied in the county agricultural extension offices and local (multi-county) water or resources management agencies. The local (multi-county) water or resources management agencies sit between farmers and other water users on the one hand, and the state resource agencies and state land grant college and experiment stations on the other hand. They are able to convene farmers, managers, and researchers for meetings and workshops on a wide range of topics. Resource district staff routinely translate farmers' needs and concerns for researchers in order to set research agendas, and they regularly help translate research results in ways that are understandable and relevant to farmers. They also translate the interests, concerns, and needs of constituents for decision makers at the state level. The district office can serve as a site for collaboration, bringing together farmers, agronomists, hydrologists, and managers to build hydrologic and agronomic models to test different policy options for water management. Finally, the district office is a place where discussions between multiple, often conflicting, perspectives can be mediated and conflict resolved. The areas with a management district that served these functions were able to integrate research on climate and hydrology with decision making and produce outcomes that reduced social and ecological vulnerability better than areas without such boundary management institutions (see Figure 1) (Cash, 2001; Cash et al., 2003).

Our research suggests that information that is co-produced through the actions of boundary organizations has three critical attributes, which have been the focus of recent

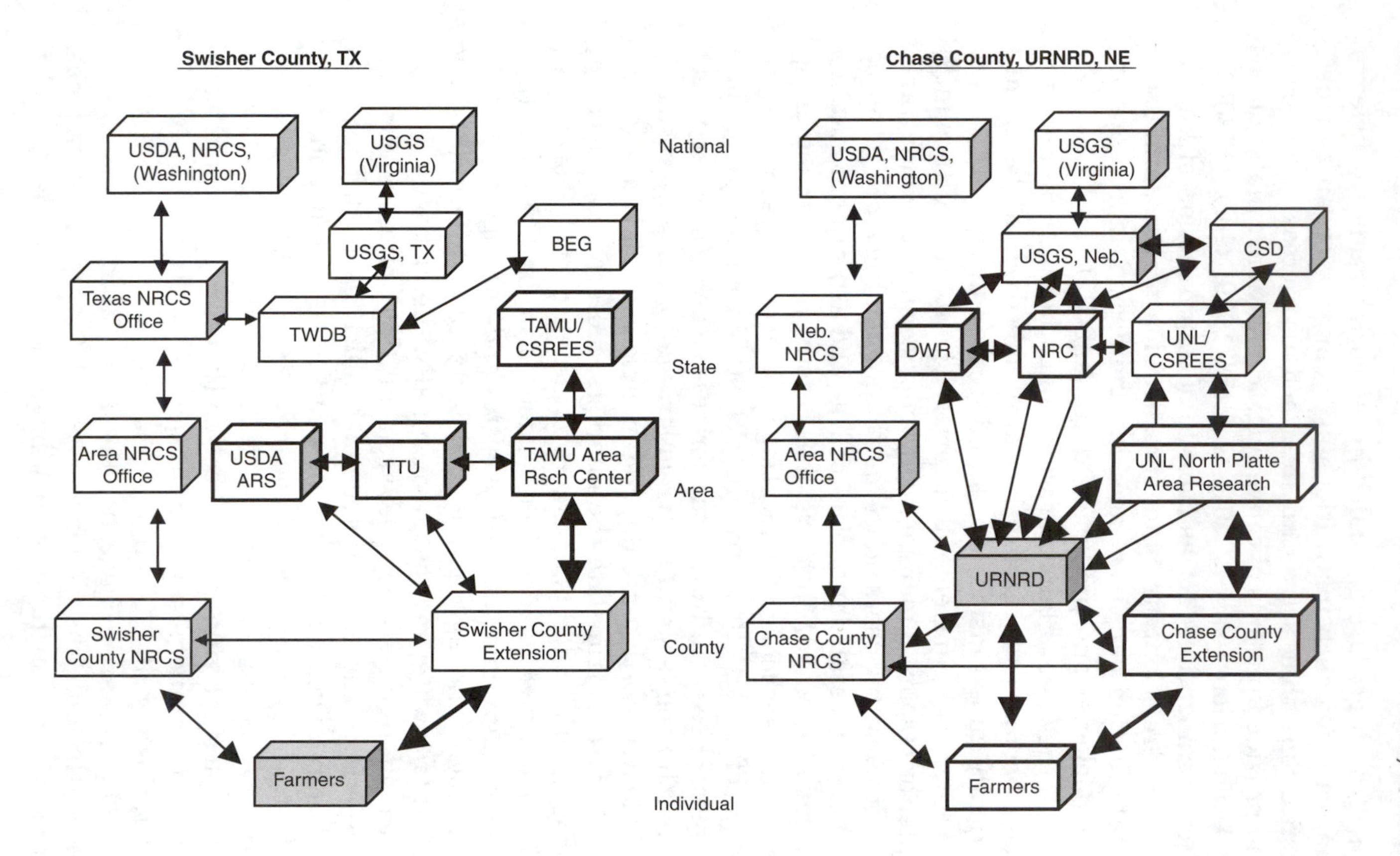
Swisher County, TX
USDA, NRCS, (Washington)
USGS (Virginia)
USGS, TX
BEG
Texas NRCS Office
TWDB
TAMU/ CSREES
Area NRCS Office
USDA ARS
TTU
TAMU Area Rsch Center
Swisher County NRCS
Swisher County Extension
Farmers
National
State
Area
County
Individual
Chase County, URNRD, NE
USDA, NRCS, (Washington)
USGS (Virginia)
USGS, Neb.
CSD
Neb. NRCS
DWR
NRC
UNL/ CSREES
Area NRCS Office
UNL North Platte Area Research
URNRD
Chase County NRCS
Chase County Extension
Farmers

Figure 1 This schematic diagram illustrates the nodes and links in the decision-making and scientific research network for two counties in the U.S. Great Plains. The system in which Swisher County, Texas, is embedded is relatively sparse, with relatively weak connections across scale and between decision-making and scientific nodes. Although the Swisher County Extension Office plays the role of a boundary organization, its capacity is somewhat limited because of the lack of connection across levels and with the policy arena. The network that includes Chase County, Nebraska, however, has a rich network that links federal scientific organizations such as the U.S. Geological Survey with the state-level and substate research institutions and links this research system with state agencies and the local management district (Upper Republican Natural Resources District [URNRD]). As is illustrated in the diagram, URNRD and the Chase County Extension Office play critical roles managing the boundaries across levels and between science and decision making. (Acronyms: ARD, Agricultural Research Division; CSD, Conservation and Survey Division of the University of Nebraska–Lincoln; CSREES, Cooperative State Research, Education, and Extension Service; DWR, Department of Water Resources (Nebraska); NRC, Natural Resources Commission (Nebraska); NRCS, Natural Resources Conservation Service; TAMU, Texas A&M University; TTU, Texas Tech University, Lubbock; TWDB, Texas Water Development Board; UNL, University of Nebraska–Lincoln; URNRD, Upper Republican Natural Resources District; USDA, U.S. Department of Agriculture; USGS, U.S. Geological Survey.)

inquiry: salience (its relevance to decision-making bodies or publics), credibility (its technical believability, or whether or not it is endorsed by relevant evaluative communities), and legitimacy (how fair an information-producing process is and whether it considers appropriate values, concerns, and perspectives of different actors) (Cash et al., 2003; Eckley et al., 2002). We refer to these three judgments as attributions because they are not objective, or even readily agreed-upon, characteristics of a knowledge production system but rather involve actor-specific judgments using different criteria and standards. Thus, salience, credibility, and legitimacy are perceived and judged differently by different audiences. Given the diverse nature of many kinds of natural resource problems such as drought, the fact multiple perceptions exist about what is salient, credible, and legitimate suggests an important connection between boundary organizations and salience, credibility, and legitimacy.

Performing functions of convening, translating, collaborating, and mediating, boundary organizations are especially well suited to helping the negotiation of information production so that the information produced is salient to a potential user and is credible, valuable to the scientist who produced it. A boundary organization, in offering a site for co-production, also can facilitate a legitimate process in which users and producers both feel they have a stake and voice at the table. In many different contexts in the Great Plains, this balancing of different needs and different perspectives has led to the production and use of hydrogeologic models; a broad network of potential evapotranspiration monitors that provide daily information on crop needs; the development of highly efficient irrigation technologies such as the nozzle systems on center-pivot systems; and many other systems that link technical knowledge derived from universities, government agencies, or industry with decision making on the ground (Cash, 2001).

In the United States, the National Drought Mitigation Center (NDMC; *http://www.drought.unl.edu*) appears to function as a boundary organization, with institutional mechanisms in place that facilitate the production of salient, credible, and legitimate information. Housed at the University of Nebraska–Lincoln (a major research institution), the mission of the NDMC is to bridge the data collection, monitoring, and research capabilities at the university and elsewhere with agencies making decisions about drought preparedness and policy. In 1998, for example, the U.S. Congress established the National Drought Policy Commission, charged with developing the framework for a massive restructuring of national drought preparedness and response efforts, and the NDMC played a central role in convening and legitimizing the final product of this commission. In serving such functions, the NDMC seeks to produce high-quality (*credible*) information about drought that is timely, useful, and relevant to decision makers' needs (*salient*) at multiple levels (local, state, national). By encouraging collaborative research and multiple avenues for two-way dialogue (e.g., workshops, conferences, etc.), it is establishing a process that has a high probability of being viewed as fair and inclusive (*legitimate*) by multiple actors from different perspectives.

Given its institutional structure, it is difficult to imagine that the NDMC would not reduce drought-related vulnerability in some ways. Yet the charge of the NDMC is not simple. Its success will be defined in part by how well it collaborates with other, more locally focused institutions also charged with mobilizing information on drought-related vulnerability. This information is almost certainly a moving target, given the spatial and temporal variability of drought and the difficulty in defining it (Wilhite, 2001). The NDMC and its partner institutions must constantly communicate with each other as well as with their end user base.

Outside the United States, the recent focus on reducing drought-related vulnerability has increasingly centered on seasonal climate forecasts (SCFs) (Dilley, 2000). The premise is straightforward: if people are suffering because of drought, and if they lack detailed foreknowledge of the drought, then improved forecasts should reduce the suffering. In the last 20 years we have seen dramatic progress in scientific efforts to understand seasonal to interannual climate dynamics, to characterize social and environmental impacts, and to predict

El Niño/Southern Oscillation (ENSO) events several months in advance. As skill in forecasting has improved, concerted efforts to reduce social vulnerability by applying this growing knowledge to decision making have been made by international organizations, national agencies, and research institutions, which serve a boundary function. ENSO application efforts have been targeted at such areas as emergency preparedness, agriculture, food security, tourism, public health, and fisheries, especially in drought-prone regions. However, the hope engendered by improvements in scientific skill has not always been realized.

Multiple challenges (unanticipated, for the most part) have impeded the integration of forecasting information into decision making. Some challenges stem from distortions and manipulation of climate information for political reasons (Broad et al., 2002; Lemos, in press; Lemos et al., 2002). Others stem from inadequate understanding in the scientific community of the needs, interests, and adaptation capacities of the end users of climate information (O'Brien and Vogel, 2003; Patt and Gwata, 2002). Still other obstacles originate in institutional constraints that inhibit the co-production of information, resulting in climate information that lacks salience, legitimacy, or credibility (Cash et al., 2003; McCarthy et al.). One of the primary lessons from reviewing these evaluations of climate forecasting systems is a questioning of the traditional notion of the primacy of producing scientifically valid information that will be simply picked up by decision makers and incorporated into decision making. Rather, this body of work suggests that institutions designed to harness science to address social vulnerability to natural phenomena require a more nuanced approach. Political and social contexts must be well understood; institutional structures must be in place to help manage the boundaries between science and decision making; and information must be perceived as multidimensional, relying not only on objective credibility, but also on its salience to decision makers' needs and the legitimacy of its production.

VI. CONCLUSIONS AND FUTURE DIRECTIONS

We return to our original question posed in Section I.

> Do past successes of adaptation to drought suggest that S&T will rise to future challenges posed by a greenhouse-induced increase in drought severity, not only in places where drought is a central planning focus, but also in places where drought receives less coordinated planning attention?

We argue that, in the United States at least, we can expect the S&T community to continue contributing to a meaningful reduction in social vulnerability to drought effects by enhancing adaptive capacity. But the associated institutions must be designed to communicate with each other and with the end users of the information and products in a substantive and iterative fashion. The real progress made in some places by the S&T community during the 20th century does not mean that the job is complete. The success of societal response to future droughts will be directly related to institutions' ability to generate S&T products that respond to an evolving set of end user needs associated with multiple stresses, at multiple levels of organization. Indeed, a recent report on preparing the United States for droughts in the 21st century suggests that renewed federal energy is required to catalyze improved preparedness efforts, but that state and local capabilities must not be compromised (National Drought Policy Commission, 2000). Such institutions must also be able to learn and communicate across traditional professional boundaries, so that the information produced remains salient, credible, and legitimate over time.

This list of normative goals expands on earlier work on the process of drought policy development (e.g., Wilhite and Easterling, 1987) and on whether drought policies should emphasize a preventive or proactive approach (e.g., Wilhite, 2001). If we want to make progress toward the institutional design elements called for here, we must first unpack the ways in which natural resource management institutions presently generate and promote the tools on which they base successful drought policies. Of course, it is easier to theorize about the need for realizing these normative goals than it is to realize the goals, especially given the fact that most of the institutions in question operate under increasing demands on limited resources. Nonetheless, a redirection of drought planning policies is needed, both in places where droughts currently contribute to serious human suffering (such as southern Africa) and in places where droughts have ceased to pose a mortal threat but continue to generate considerable damage (such as the United States). We believe that if S&T institutions adapt their form and function to suit end users' evolving needs, then they will be in a position to serve those needs better.

The ability of S&T institutions to respond to a strengthened drought regime under climate change remains an open question that demands further research. The answer will almost certainly vary from place to place. The specific characteristics required of a given institution depend on the set of stressors important there. In other words, drought is not the only stress people are battling (O'Brien and Leichenko, 2000). For example, S&T institutions will likely need to serve different user needs in places where climate change coincides with economic liberalization policies that are restructuring agriculture (e.g., South Africa, India) than in places where such nonclimate factors are less important (e.g., the United States, France). It should be noted that taking a place-based approach to studying global change vulnerability complicates the analysis. In particular, understanding the details of an institutional landscape is difficult because these institutions are constantly evolving. A case in point is the National Drought Policy Commission in the United States. The specific fruits of this commission are unclear at present, but it is certain to result in the building of institutions that span multiple disciplines, address multiple levels of organization, and serve multiple (possibly competing) constituencies. We hypothesize that the effectiveness of these institutions will be positively related to the extent to which they satisfy the design elements outlined in this chapter. Given the imperative to reduce the rising toll of drought impacts, testing this hypothesis in the coming years as the policies materialize will illuminate pathways for reducing those impacts, and, by extension, overall social vulnerability.

REFERENCES

Adams, WM. How beautiful is small? Scale, control and success in Kenyan irrigation. *World Development* 18(10):1309–1323, 1990.

Adger, WN; Kelly, PM. Social vulnerability to climate change and the architecture of entitlements. *Mitigation and Adaptation Strategies for Global Change* 4(3–4):253–266, 1999.

Andrews, CJ. *Humble Analysis: The Practice of Joint Fact-Finding.* Westport, CT: Praeger, 2002.

Avalos, M; DeYoung, T. Preferences for water policy in the Ogallala region of New Mexico: Distributive vs. regulatory solutions. *Policy Studies Journal* 23(4):668–685, 1995.

Berkes, F. Cross-scale institutional linkages: Perspectives from the bottom up. In: EU Weber, Ed. *The Drama of the Commons* (pp. 293–319). Washington, D.C.: National Academy Press, 2002.

Blomquist, W. *Dividing the Waters: Governing Groundwater in Southern California*. San Francisco: ICS Press, 1992.

Böhle, HG; Downing, TE; Watts, M. Climate change and social vulnerability: Toward a sociology and geography of food insecurity. *Global Environmental Change* 4(1):37–48, 1994.

Bowden, MJ; Kates, RW; Kay, PA; Riebsame, WE; Warrick, R; Johnson, DL; Gould, HA; Weiner, D. The effect of climate fluctuations on human populations: Two hypotheses. In: TM Wigley et al., Eds. *Climate and History: Studies in Past Climates and Their Impact on Man.* Cambridge: Cambridge University Press, 1981.

Broad, K; Pfaff, A; Glantz, M. Effective and equitable dissemination of seasonal-to-interannual climate forecasts: Policy implications from the Peruvian fishery during El Niño 1997–98. *Climatic Change* 54:415–438, 2002.

Bruggink, TH. Privatization versus groundwater central management: Public policy choices to prevent a water crisis in the 1990s. *American Journal of Economics and Sociology* 51(2):205–222, 1992.

Burton, I; Kates, RW; White, GF; Eds. *The Environment as Hazard.* New York: Oxford, 1978.

Cash, DW. "In order to aid in diffusing useful and practical information": Agricultural extension and boundary organizations. *Science, Technology, and Human Values* 26(4):431–453, 2001.

Cash, DW; Moser, S. Linking global and local scales: Designing dynamic assessment and management processes. *Global Environmental Change* 10(2):109–120, 2000.

Cash, DW; Clark, WC; Alcock, F; Dickson, NM; Eckley, N; Guston, D; Jäger, J; Mitchell, R. Knowledge systems for sustainable development. *Proceedings of the National Academy of Sciences* 100:8086–8091, 2003.

Clark, WC; Dickson, NM. Sustainability science: The emerging research program. *Proceedings of the National Academy of Sciences* 100(14):8059–8061, 2003.

Clark, W; Mitchell, R; Cash, DW; and Alcock, F; (forthcoming) Information as influence: how institutions mediate the impact of scientific assessments on global environmental affairs in *Global Environmental Assessments: information, Institutions, and Influence.* R; Mitchell, W; Clark, DW; Cash and F; Alcock, eds. Cambridge, MA: MIT Press.

Cutter, S. Vulnerability to environmental hazards. *Progress in Human Geography* 20(4):529–539, 1996.

Dilley, M. Reducing vulnerability to climate variability in southern Africa: The growing role of climate information. *Climatic Change* 45(1):63–73, 2000.

Downing, TE. Vulnerability to hunger in Africa: A climate change perspective. *Global Environmental Change* 1:365–380, 1991.

Downing, TE. Human dimensions research: Toward a vulnerability science? *International Human Dimensions Program Update* No. 3, 2000, 16–17.

Easterling, WE and Polsky, C; 2004. Crossing the complex divide: Linking scales for understanding coupled human–environment systems. In: R. McMaster and E. Sheppard, Eds., *Scale and Geographic Inquiry.* Blackwell, Oxford, pp. 55–64. (Also online at http://www.indp.uni-bonn.de/html/publications/update/IHDpupdate00.03.html)

Eckley, N; Clark, WC; Farrell, A; Jäger, J; Stanners, D. Designing effective assessments: The role of participation, science and governance, and focus. Global Environmental Assessment Project and the European Environment Agency, Copenhagen, Denmark, *http://www.environment/gea/pubs/gea%2Deea01ws.html?AUVAL%3Dne&INDEX%3Dgea/pubsbyauthor.html*, 2002.

The Economist. A growing thirst. January 25, p. 34, 2003.

Emel, J; Roberts, R. Changes in form and function of property rights institutions under threatened resource scarcity. *Annals of the Association of American Geographers* 78(2):241–252, 1988.

Esty, DC. 1996. Revitalizing environmental federalism. *Michigan Law Review* 95(3):570–653, 1996.

Finan, T; West, C; Austin, D; McGuire, T. Processes of adaptation to climate variability: A case study from the US Southwest. *Climate Research* 21(3):299–310, 2002.

Folke, C; Carpenter, S; Elmquist, T; Gunderson, L; Holling, CS; Walker, B; Bengtsson, J; Berkes, F; Colding, J; Danell, K; Falkenmark, M; Gordon, L; Kasperson, R; Kautsky, N; Kinzig, A; Levin, S; Mäler, K-G; Moberg, F; Ohlsson, L; Olsson, P; Ostrom, E; Reid, W; Rockström, J; Savenije, H; Svedin, U. Resilience and sustainable development: Building adaptive capacity in a world of transformations, ICSU, Paris, France, *http://www.resalliance.org/reports/resilience_and_sustainable_development.pdf*, 2002.

Gieryn, TF. Boundaries of science. In: S Jasanoff et al., Eds. *Handbook of Science and Technology Studies.* Thousand Oaks, CA: Sage Publications, 1995.

Glantz, MH. Drought, desertification, and food production. In: MH Glantz, Ed. *Drought Follows the Plow: Cultivating Marginal Areas.* London: Cambridge University Press, 1994.

Glantz, MH; Ausubel, JH. Impact assessment by analogy: Comparing the impacts of the Ogallala Aquifer depletion and CO_2-induced climate change. In: MH Glantz, Ed. *Societal Responses to Regional Climatic Change: Forecasting by Analogy* (pp. 113–142). Boulder, CO: Westview Press, 1988.

Gunderson, LH; Holling, CS; Light, SS; Eds. *Barriers and Bridges to the Renewal of Ecosystems and Institutions.* New York: Columbia University Press, 1995.

Guston, DH. Stabilizing the boundary between politics and science: The role of the Office of Technology Transfer as a boundary organization. *Social Studies of Science* 29(1):87–112, 1999.

Guston, DH. Boundary organizations in environmental policy and science: An introduction. *Science, Technology, and Human Values* 26(4):399–408, 2001.

Hanemann, WM. Adaptation and its measurement: An editorial comment. *Climatic Change* 45:571–581, 2000.

Holling, CS, Ed. *Adaptive Environmental Assessment and Management. International Series on Applied Systems Analysis.* New York: Wiley & Sons, 1978.

Hulett, GK. The future of the grasslands. In: MP Lawson, ME Baker, Eds. *The Great Plains: Perspectives and Prospects.* Lincoln, NE: University of Nebraska Press, 1981.

Jasanoff, SS. Contested boundaries in policy-relevant science. *Social Studies of Science* 17:195–230, 1987.

Kasperson, RE; Renn, O; Slovic, P; Brown, H; Emel, J; Goble, R; Kasperson, JX; Ratick, S. The social amplification of risk: A conceptual framework. *Risk Analysis* 8(2):177–187, 1988.

Kates, RW. The interaction of climate and society. In: RW Kates, et al., Eds. *Climate Impact Assessment: Studies of the Interaction of Climate and Society.* Chichester, U.K.: Wiley, 1985.

Kates, RW; Clark, WC; Corell, R; Hall, JM; Jaeger, CC; Lowe, I; McCarthy, JJ; Schellnhuber, HJ; Bolin, B; Dickson, NM; Faucheux, S; Gallopin, GC; Gruebler, A; Huntley, B; Jäger, J; Jodha, NS; Kasperson, RE; Mabogunje, A; Matson, P; Mooney, H; Moore, B III; O'Riordan, T; Svedin, U. Sustainability science. *Science* 292 (April 27):641–642, 2001.

Kelly, PM; Adger, WN. Theory and practice in assessing vulnerability to climate change and facilitating adaptation. *Climatic Change* 47:325–352, 2000.

Keohane, RO; Levy, MA; Eds. *Institutions for Environmental Aid.* Cambridge, MA: MIT Press, 1996.

Keohane, RO; Nye, JS. *Power and Interdependence.* Glenview, IL: Scott, Foresman, 1989.

Keohane, RO; Ostrom, E; Eds. *Local Commons and Global Interdependence: Heterogeneity and Cooperation in Two Domains.* London: Sage Publications, 1995.

Keohane, RO; Haas, PM; Levy, MA. The effectiveness of international environmental institutions. In: MA Levy, Ed. *Institutions for the Earth*. Cambridge, MA: MIT Press, 1994.

Kincaid, J. Intergovernmental costs and coordination in U.S. environmental protection. In: B Galligan, Ed. *Federalism and the Environment: Environmental Policymaking in Australia, Canada, and the United States* (pp. 79–102). Westport, CT: Greenwood Press, 1996.

Kromm, DE; White, SE. *Conserving Water in the High Plains*. Manhattan, KS: Kansas State University, 1990.

Lemos, MC. A tale of two policies: The Politics of Seasonal Climate Forecast Use in Ceara, Brazil. *Policy Sciences*. Vol.32, 2, pp. 101–123.

Lemos, MC; Finan, TJ; Fox, RW; Nelson, DR; Tucker, J. The use of seasonal climate forecasting in policymaking: Lessons from Northeast Brazil. *Climatic Change* 55:479–507, 2002.

Levin, SA. Management and the problem of scale. *Conservation Ecology* 1 (1):13, http://www.consecol.org/Journal/vol1/iss1/art13, 1997.

Liverman, D. Vulnerability to global environmental change. In: JX Kasperson, RE Kasperson, Eds. *Global Environmental Risk* (pp. 201–216). Tokyo: United Nations University Press, 2001.

McCarthy, JJ; Canziani, OF; Leary, NA; Dokken, DJ; White, KS; Eds. *Climate Change 2001: Impacts, Adaptation, and Vulnerability. Intergovernmental Panel on Climate Change*. London: Cambridge University Press, 2001.

McCarthy, JJ; Martello, ML; Corell, R; Eckley, N; Fox, S; Hovelsrud-Broda, G; Mathiesen, S; Polsky, C; Selin, H; Tyler, N; Bull, KS; Eira, IMG; Eira, NI; Eriksen, S; Hanssen-Bauer, I; Kalstad, JK; Nellemann, C; Oskal, N; Reinert, ES; Siegel-Causey, D; Storeheier, PV and Turi, JM; 2004. Climate Change in the Context of Multiple Stressors and Resilience. In: AMAP Eds., Impacts of a Warming Climate - Arctic Climate Impact Assessment. *Arctic Monitoring and Assessment Program*. Carmbridge University Press, Cambridge, UK, pp. 140.

McGinnis, M, Ed. *Polycentric Governance and Development: Readings from the Workshop in Political Theory and Policy Analysis. Institutional Analysis*. Ann Arbor, MI: The University of Michigan Press, 1999.

McGuire, VL. Water-Level Changes in the High Plains Aquifer, Predevelopment to 2001, 1999 to 2000, and 2000 to 2001. U.S. Geological Survey, *http://water.usgs.gov/pubs/fs/FS078-03/*, 2003.

Mendelsohn, R; Nordhaus, W; Shaw, D. The impact of global warming on agriculture: A Ricardian analysis. *American Economic Review* 84(4):753–771, 1994.

Mitchell, R; Bernauer, T. Empirical research on international environmental policy: Designing qualitative case studies. *Journal of Environment and Development* 7(1):4–31, 1998.

National Drought Policy Commission. Preparing for Drought in the 21st Century—Report of the National Drought Policy Commission, United States Department of Agriculture, Washington, D.C., *http://www.fsa.usda.gov/drought/finalreport/reports.htm*, 2000.

National Research Council. *Our Common Journey: A Transition toward Sustainability.* Board on Sustainable Development. Washington, D.C.: National Academy Press, 1999.

Nye, JSJ; Donahue, JD; Eds. *Governance in a Globalizing World.* Washington, D.C.: Brookings Institution Press, 2000.

O'Brien, K; Leichenko, R. Double exposure: Assessing the impacts of climate change within the context of economic globalization. *Global Environmental Change* 10:221–232, 2000.

O'Brien, K; Vogel, C; Eds. *Coping with Climate Variability: The Use of Seasonal Climate Forecasts in Southern Africa. Ashgate Series in Environmental Policy and Practice.* Abingdon, U.K.: Ashgate, 2003.

O'Neill, RV. Hierarchy theory and global change. In: T. Rosswall et al., Eds. *SCOPE 35, Scales and Global Change: Spatial and Temporal Variability in Biospheric and Geospheric Processes* (pp. 29–45). Chichester, U.K.: John Wiley and Sons, 1988.

Opie, J. Moral geography in High Plains history. *The Geographical Review* 88(2):241–258, 1998.

Opie, J. *Ogallala: Water for a Dry Land.* Lincoln, NE: University of Nebraska Press, 2000.

Ostrom, E. *Governing the Commons: The Evolution of Institutions for Collective Action.* Cambridge, U.K.: Cambridge University Press, 1990.

Ostrom, E. Scales, polycentricity, and incentives: Designing complexity to govern complexity. In: JA McNeely, Ed. *Protection of Biodiversity: Converging Strategies* (pp. 149–167). Durham, NC: Duke University Press, 1998.

Parry, ML. Viewpoint—Climate change: Where should our research priorities be? *Global Environmental Change* 11:257–260, 2001.

Patt, AG; Gwata, C. Effective seasonal climate forecast applications: Examining constraints for subsistence farmers in Zimbabwe. *Global Environmental Change* 12:185–195, 2002.

Polsky, C. 2004. Putting space and time in Ricardian climate change impact studies: The case of agriculture in the U.S. Great Plains. *Annals of the Association of American Geographers*, 94(3): 549–564.

Polsky, C; Schröter, D; Patt, A; Gaffin, S; Martello, ML; Neff, R; Pulsipher, A; Selin, H. Assessing vulnerabilities to the effects of global change: An eight-step approach. Research and Assessment Systems for Sustainability Program Discussion Paper 2003–05, Environment and Natural Resources Program, Belfer Center for Science and International Affairs, Kennedy School of Government, Harvard University, Cambridge, MA, *http://ksgnotes1.harvard.edu/BCSIA/sust.nsf/pubs/pub75*, 2003.

Popper, DE; Popper, FJ. The Great Plains: From dust to dust. *Planning* (December): 13–18, 1987.

Riebsame, WE. Managing drought impacts on agriculture: The Great Plains experience. In: RH Platt, G Macinko, Eds. *Beyond the Urban Fringe: Land Use Issues of Nonmetropolitan America* (pp. 257–270). Minneapolis, MN: University of Minnesota Press, 1983.

Riebsame, WE. The United States Great Plains. In: BL Turner et al., Eds. *The Earth as Transformed by Human Action: Global and Regional Changes in the Biosphere over the Past 300 Years* (pp. 561–576). Cambridge: Cambridge University Press, 1990.

Riebsame, WE. Sustainability of the Great Plains in an uncertain climate. *Great Plains Research* 1(1):133–151, 1991.

Riebsame, WE. The historical bias for growth and intensification in Great Plains agricultural development. In: AW Gilg, Ed. *Progress in Rural Policy and Planning*. New York: John Wiley and Sons, 1994.

Roberts, R; Emel, J. The Llano Estacado of the American Southern High Plains. In: JX Kasperson, RE Kasperson, BL Turner II, Eds. *Regions at Risk: Comparisons of Threatened Environments* (pp. 255–303). Tokyo: United Nations University Press, 1995.

Rosenberg, NJ; Wilhite, DA. Proactive or risk management cases: Drought in the U.S. Great Plains. In: V Yevjevich et al., Eds. *Coping with Droughts.* Chelsea, MI: BookCrafters, 1983.

Sachs, JD. *Macroeconomics and Health: Investing in Health for Economic Development.* Geneva: World Health Organization, *http://www.cid.harvard.edu/cidcmh/CMHReport.pdf*, 2001.

Schröter, Dagmar, Colin Polsky, and Anthony G. Patt, 2005 (in press). "Assessing Vulnerabilities to the Effects of Global Change: An Eight Step Approach" *Mitigationa and Adaptation Stratiegies for Global Change.*

Simon, HA. 1962. The architecture of complexity. *Proceedings of the American Philosophical Society* 106(6):467–482, 1962.

Somma, M. Local autonomy and groundwater district formation in High-Plains West Texas. *Publius* 24:1–10, 1994.

Taylor, JG; Downton, MW; Stewart, TR. Adapting to environmental change: Perceptions and farming practices in the Ogallala Aquifer region. In: EE Whitehead et al., Eds. *Arid Lands: Today and Tomorrow, An International Research and Development Conference* (pp. 665–684). Tucson, AZ: Westview Press, 1988.

Turner, BL; Kasperson, RE; Matson, P; McCarthy, JJ; Corell, RW; Christensen, L; Eckley, N; Kasperson, JX; Luers, A; Martello, ML; Polsky, C; Pulsipher, A; Schiller, A. A framework for vulnerability analysis in sustainability science. *Proceedings of the National Academy of Sciences* 100(14):8080–8085, 2003.

Warrick, R; Bowden, M. The changing impacts of droughts in the Great Plains. In: MP Lawson, ME Baker, Eds. *The Great Plains: Perspectives and Prospects.* Lincoln, NE: University of Nebraska Press, 1981.

Webb, WP. *The Great Plains.* Lincoln, NE: University of Nebraska Press, 1931.

White, SE. Ogallala oases: Water use, population redistribution, and policy implications in the High Plains of Western Kansas, 1980–1990. *Annals of the Association of American Geographers* 84(1):29–45, 1994.

Wilbanks, TJ; Kates, RW. Global change in local places: How scale matters. *Climatic Change* 43(3):601–628, 1999.

Wilhite, DA. Government response to drought in the United States with particular reference to the Great Plains. *Journal of Climate and Applied Meteorology* 22(1):40–50, 1983.

Wilhite, DA. The Ogallala Aquifer and carbon dioxide: Are policy responses applicable? In: MH Glantz, Ed. *Societal Responses to Regional Climatic Change: Forecasting by Analogy* (pp. 353–374). Boulder, CO: Westview Press, 1988.

Wilhite, DA. Moving beyond crisis management. *Forum for Applied Research and Public Policy* 16(1):20–28, 2001.

Wilhite, DA; Easterling, WE. Drought policy: Toward a plan of action. In: DA Wilhite, WE Easterling, Eds. *Planning for Drought: Toward a Reduction of Societal Vulnerability* (pp. 573–583). Boulder, CO: Westview Press, 1987.

Wilhite, DA; Wood, DA. Revisiting drought relief and management efforts in the West: Have we learned from the past? *Journal of the West* 40(3):18–25, 2001.

World Bank. *World Development Report 1999/2000: Entering the 21st Century.* New York: Oxford University Press, 1999.

Young, OR. The problem of scale in human/environment relationships. In: E Ostrom, Ed. *Local Commons and Global Interdependence: Heterogeneity and Cooperation in Two Domains* (pp. 27–45). London: Sage Publications, 1995.

Young, OR, Ed. Science Plan for the Project on the Institutional Dimensions of Global Environmental Change. Bonn: International Human Dimensions Programme on Global Environmental Change, 1999.

Young, OR. Institutional interplay: The environmental consequences of cross-scale interactions. In: EU Weber, Ed. *The Drama of the Commons* (pp. 259–291). Washington, D.C.: National Academy Press, 2002.

Part III

Case Studies in Drought and Water Management: The Role of Science and Technology

10

The Hardest Working River: Drought and Critical Water Problems in the Colorado River Basin

ROGER S. PULWARTY, KATHERINE L. JACOBS, AND RANDALL M. DOLE

CONTENTS

> You are piling up a heritage of conflict and litigation over water rights for there is not sufficient water to supply the land ...
>
> **John Wesley Powell, 1893**
> *International Irrigation Conference, Los Angeles*
> *cited in Stegner, 1954, p. 343*

I. INTRODUCTION: HISTORY OF COLORADO RIVER BASIN DEVELOPMENT

The Colorado River flows 2300 km (about 1400 mi) from the high mountain regions of Colorado through seven basin states to the Sea of Cortez in Mexico (Figure 1). The river supplies much of the water needs of seven U.S. states, two Mexican states, and 34 Native American tribes. These represent a population of 25 million inhabitants, with a projection of 38 million by the year 2020. Approximately 2% of the basin is in Mexico. The Colorado does not discharge a large volume of water. Because of the scale of impoundments and withdrawals relative to its flow, the Colorado has been called the most legislated and managed river in the world. It has also been called the most "cussed" and "discussed" river in the United States. About 86% of the Colorado's annual runoff originates within only 15% of the area, in the high mountains of Colorado and the Wind River Range in Wyoming. In the semiarid Southwest, even relatively small changes in precipitation can have

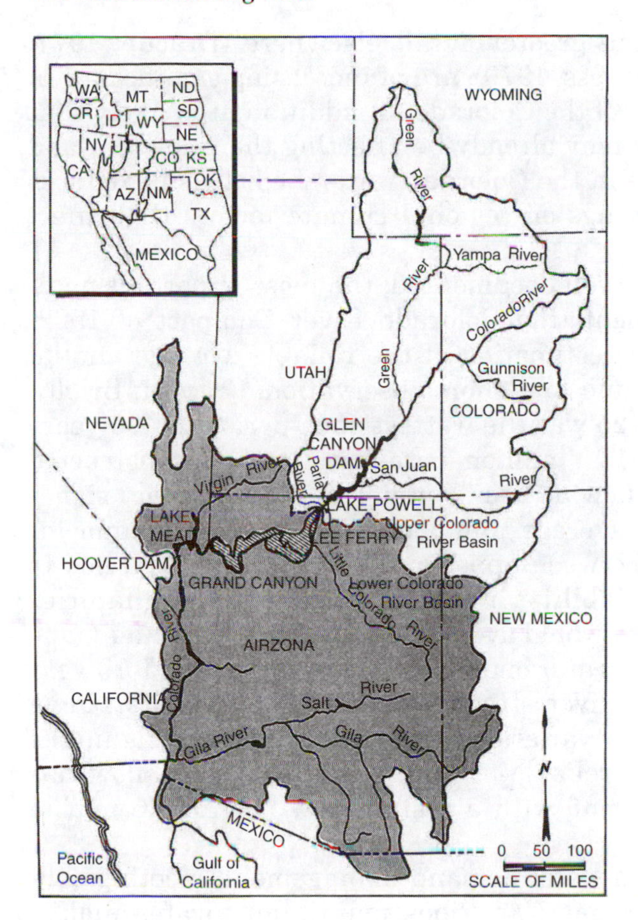

Figure 1 The Colorado River basin. (From the U.S. Department of the Interior, Bureau of Reclamation.)

large impacts on water supplies. The coefficient of variation for the Colorado is about 33%.

Climate and weather events form a variable background on which water agreements and conflicts are played out. Indeed, Powell's comment above, as dire as it might seem, was not made in the context of potentially large swings in the climate system. The specter of long-term climate variations overlays a series of other issues, including growth in municipal and industrial water demands, groundwater depletion, unmet ecosystem needs, and water quality requirements. Decadal-scale climatic factors influencing present water allocations, discussed in greater detail elsewhere (Dracup, 1977; Stockton and Boggess, 1979), are of increasing significance in the management of the Colorado. In addition, it is likely that climatic changes may already be affecting the snowpack and runoff conditions in the Colorado watershed. This introduces a new set of forcings on regional climate factors that affect water supply.

As has been well documented, the most important management agreement (the Colorado River Compact of 1922) was based on overestimation of the reliable average annual supply of water due to a short observational record. Briefly, the period 1905–25 was the wettest such period in 400 years of record, with 16.4 million acre-feet (maf[1]) reconstructed annual average flow at Lees Ferry. The 1922 compact signatories used this average number as the base minimum for fixed allocation between upper and lower basins. As a nod to interannual variability in water supply, the signatories assumed that flow would average out over 10 years and made the downstream requirement 75 maf over the said 10-year period. Colorado River streamflow, however, exhibits strong decadal and longer variations (Figure 2). Since the signing of the compact, the reliable estimated annual virgin flow has been about 14.3 maf, with a historic low flow of 5.6 maf in 1934.

Emphases on water demand management, meeting obligations to Native American tribes, maintaining water quality, and environmental concerns have also altered the traditional roles of federal, state, and local agencies. The impacts of recent events such as the continuing regional-scale droughts since 1999, including the extreme drought of 2002, and recent enforcements restricting California to its compact allotment are only just beginning to be understood in terms of system criticality and requirements for noncrisis or proactive mitigation of drought impacts.

---

[1] 1 maf = 1.24 million liters (325, 851 million gallons). Million acre-feet (maf) is used as the unit of water volume throughout this chapter. All entities on the Colorado River use maf as the unit of measure.

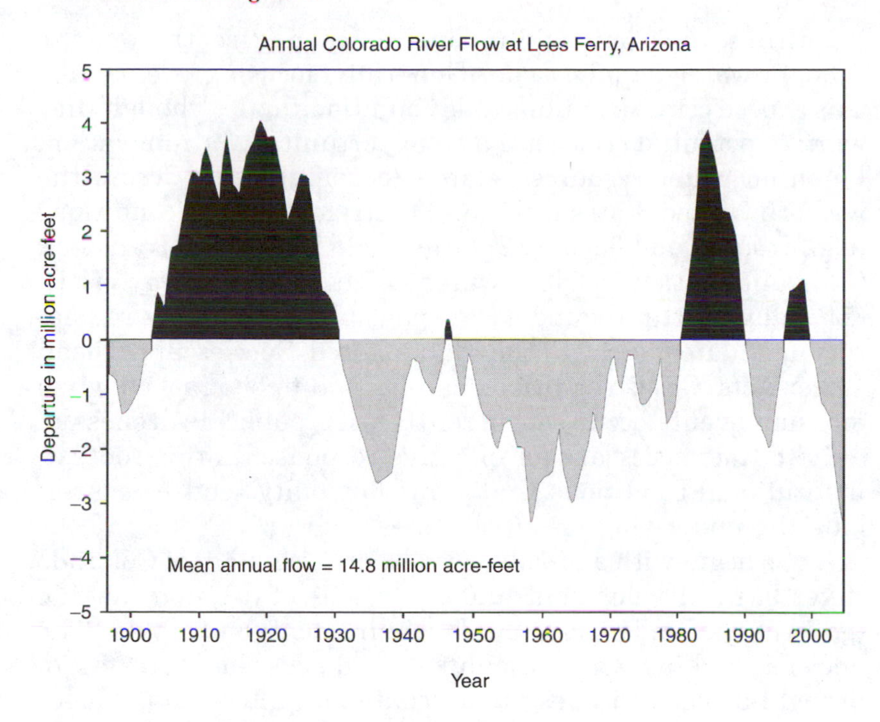

Figure 2 Decadal-scale variability of Colorado River streamflow at Lees Ferry, 1896–2003. Smoothed using a 9-year moving average. (Data from the U.S. Department of the Interior, Bureau of Reclamation.)

This chapter uses climate-sensitive decision environments along the Colorado River to illustrate the breadth and complexity of the water management issues and the role of climate in these contexts. The four examples are in: (1) the border region: international issues; (2) Arizona and California: interstate issues in the Lower Basin; (3) Native American water rights; and (4) conjunctive use and management: groundwater and surface water in Arizona.

Recent drought impacts on the Colorado River reservoirs have raised significant concern about the reliability of deliveries in the event of a decadal or multi-decadal drought. Until recently, the expectation of Colorado River managers was that significant shortages in the Lower Basin would not occur until after 2030. With reservoir levels at historic lows, newspaper headlines and politicians are focused directly on the drought/water supply issue. Generally, focusing events like this expose critically vulnerable conditions and, although they warn of potential crisis, are also opportunities for innovation. Potential water resource–related focusing events across the western United States include: (1) extreme climatic conditions (e.g., drought and floods); (2) large-scale inter-basin transfers; (3) quantification of tribal water rights; (4) an energy crisis; (5) changing transboundary responsibilities; and (6) regulatory mandates such as the Endangered Species and Clean Water Acts. Crisis conditions can be said to be reached when focusing events occur concurrently with public awareness of a finite time necessary for effective response. In this context, institutional conditions that limit flexibility tend to exacerbate the underlying resource issues.

We begin with a broad overview of the history of Colorado River basin development and the scales of decision making (governance and operational requirements) involved. The decision-making environments are discussed in terms of drought-sensitive issues at international, inter-state, Native American, and state levels. The development of the Colorado River Compact (and its use of a limited record of streamflow) mentioned above is discussed in great detail in numerous books and articles (see Weatherford and Brown, 1986) and will be referred to here only when it introduces a criticality to the management problem being considered. Two issues that were not in the original compact but have since become more important will be addressed in some detail: conjunctive use (i.e., joint use of surface and groundwater) and water quality.

II. SOCIAL AND ECONOMIC CONTEXTS

Demographic, legal, and environmental changes can and have disrupted existing relationships and current perspectives about the interactions among society, climate, and water. Nowhere is this more apparent than in the many transboundary situations that dominate Colorado River management. The Colorado River has been the subject of extensive negotiations and litigation. The federal government accounts for

56% of the land within the basin; Indian reservations, 16.5%; states, 8.5%; and private ownership, 19% (Weatherford and Brown, 1986). As a result, a complex set of federal laws, compacts, court decisions, treaties, state laws, and other agreements collectively known as the "Law of the River" has been developed (Table 1). These play out in terms of interstate agreements (e.g., the Colorado River Compact) and transnational (U.S.–Mexico) settings. A study by an alliance of seven western water resources institutes (Powell Consortium, 1995) offers the following counterintuitive result: Although the Lower Colorado River Basin within the United States is indeed drier than the Upper Basin, it is the Upper Basin that is vulnerable to severe, long-term climatological drought because of the 1922 agreement to provide a fixed amount of water to the Lower Basin. However, the Lower Basin is subject to water supply limitations brought on by growth and inflexible allocation arrangements. This unprecedented growth has occurred during a wetter-than-average 25-year period (1975–99), which may have resulted in some degree of complacency about water availability.

The chronology in Table 1 reflects the changing values of water rights in the new West based on tourism and recreational economies. Management has evolved from two classic approaches to integrated river basin development: (1) large-scale investments in water projects integrating economic and engineering objectives, and (2) negotiation of inter-state and international agreements for the management of shared resources.

Recently, emphases have shifted to integration of irrigation with other agricultural land uses, wastewater reuse, and conjunctive management of ground and surface water systems. Most important are the trends toward public involvement and participation in decision-making processes and the incorporation of institutional and behavioral considerations in the planning and implementation processes.

Frederick et al. (1996) concluded that in the upper Colorado region the value of water for recreation, fish, and wildlife was US$51 per acre-foot, compared to US$21 for hydropower and US$5 for irrigation. Even given the limited

TABLE 1 The Colorado River: Relevant Events and Agreements, 1902–2004.

Year	Event
1902	Arthur Powell Davis, USGS engineer (future head of the Bureau of Reclamation), proposes "the gradual comprehensive development of the Colorado by a series of large storage reservoirs."
1905	Flood waters break into Imperial Valley, creating the Salton Sea over 2 years.
1919	Kettner Bill authorizes building of aqueduct.
1920	Kincaid Act authorizes data gathering for the All-American Canal. Population of Los Angeles reaches 600,000 (600% more than in 1900). Mulholland and Scattergood endorse Davis's plan to use Colorado to meet "all future electricity needs." Denver population reaches 260,000 (100% increase since 1900).
1922*	Colorado River Compact. Upper and Lower Basins demarcated at Lees Ferry. All basin states except Arizona ratify agreement. Indian rights considered "negligible."
1923	Dry year. Los Angeles looks to Colorado for water as well as electricity.
1927	Metropolitan Water District of Southern California approved by state legislature.
1928*	Boulder Canyon Act (BCA) approved in Congress. Authorizes construction of Hoover Dam. 1922 compact ratified. Lower Basin allotments apportioned.
1930	Arizona v. California. Arizona requests that the BCA be declared unconstitutional.
1931*	California Seven Party Agreement on municipal vs. agricultural use
1935	Hoover Dam completed. California purchases all power produced.
1944*	Colorado River Compact ratified by Arizona.
1945	Mexican Treaty approved in Congress, with support from Upper Basin, Arizona, and Texas. Mexico receives 1.5 maf despite objections from California.
1948*	Upper Basin Compact: Allots Colorado 51.75%, Utah 23%, Wyoming 14%, New Mexico 11.25% (and 50,000 af to Arizona above Lees Ferry).
1956*	Colorado River Storage Project Act. *Arizona v. California.*
1963	Glen Canyon Dam completed. Lake Powell begins filling. Indian uses charged against the state in which a reservation was located.
1964	*Arizona v. California* Supreme Court decision. Settles 25-year dispute. Allows Arizona's decision to build the Central Arizona Project (CAP) to fully use its allotment.

1968*	Colorado River Basin Project Act. Construction of major water developments in both Upper and Lower Basins. CAP designated junior right.
1970*	Criteria for Coordinated Long-Range Operation of Colorado River System. Glen Canyon Dam releases to maintain balance between Lake Powell and Lake Mead.
1973*	Minute No. 242 of the U.S.–Mexico International Boundary Commission.
1974*	Colorado River Basin Salinity Control Act. Authorized desalination and salinity control projects (including Yuma Desal Plant).
1987	Increased generator capacity and resulting changes in operations require environmental impact statement (EIS) for Glen Canyon Dam.
1994	Draft EIS issued. U.S. Fish & Wildlife Service BiOp on Glen Canyon operations.
1996	Controlled flood released from Glen Canyon Dam.
2001	Colorado River Interim Surplus Guidelines. Surplus in Lower Basin to be divided between California and Arizona. Quantification Settlement Agreement.
2004	Worst drought period in 100 years continues (since 1999).

Note: Asterisked (*) years denote passage of principal documents forming the "Law of the River."

reliability of the precision of these numbers, they reflect changing values of water rights in the new West based on tourism and recreation. Booker and Young (1994) concluded that efficient administration would require a large reallocation from the Upper Basin to the Lower Basin to reflect the low marginal values of irrigation water in the Upper Basin and the high instream values generated between the two basins. Efficiency is obviously not the only criterion for management of a multifaceted and socially constrained resource such as water. In the case of the Colorado it has become virtually impossible to answer the question "Who manages this basin?" (even with the Secretary of the Interior designated as "water master" for the Lower Basin) without listing dozens of government agencies, legal and diplomatic instruments and precedents, private-sector interests, and community-based interests (Varady et al., 2001). Climate-sensitive decisions in the Colorado River basin thus involve and cross the many temporal and spatial scales through which water of varying quantity and quality flows (Pulwarty and Melis, 2001).

A. Water Quantity

As a result of climatological droughts experienced during the 1930s, 1950s, and 1970s, the Colorado system as a whole is operated to maximize the amount of water in storage for protection against dry years. The full Colorado reservoir system stores about four times the annual flow. Lake Mead and Lake Powell are the two largest man-made lakes in the United States. Under the Colorado River Compact and subsequent international treaties, 7.5 maf are allocated to the four Upper Basin states of Colorado, Utah, Wyoming, and New Mexico; 7.5 maf to the three Lower Basin states of Arizona (2.8 maf), Nevada (0.3 maf), and California (4.4 maf); and 1.5 maf to Mexico. At present the estimated use within the Lower Basin is 8.0 maf (including return flows but not including the Mexican requirement), whereas for the Upper Basin use is estimated for 1996–2000 at 4.5–5.0 maf (Bureau of Reclamation, 2001). As such, the main focus of this chapter is on Lower Basin problems and innovations. However, in the context of severe, sustained drought, the Upper Basin could experience significant shortfalls as a result of the compact requirements to maintain the flows into Lake Powell. A "compact call" could limit diversions that currently serve multiple users in Colorado, Utah, and New Mexico.

Approximately 80% of the river's supply is used for agriculture. The largest user of agricultural water is the Imperial Irrigation District (IID) in southern California, which alone accounts for approximately 2.87 maf annually (1964–96 average), or almost 20% of the river's average annual flow. Even without the pressure of the ongoing drought, usage trends were approaching system criticality (Figure 3). The California Department of Water Resources estimates that, because of population pressure, California will face shortfalls of 4–9 maf per year by 2020. Planners in Nevada anticipate a population growth from 1.8 million in 2000 to 3.5 million by 2020. Southern Nevada, which includes Las Vegas, is now one of the fastest-growing urban areas in the country and is expected to fully utilize its basic apportionment by 2010. An earlier estimate was for this point to be reached by 2030. Water use in Utah is anticipated to almost triple over the next 50 years, from 645,000 af in 2000 to 1,695,000 af in 2050. By that time the state will be facing a projected water shortage of an estimated 186,000 af even though conservation and conversion of water use by agriculture will contribute 783,000 af of savings (see Morrison et al., 1996; Pontius, 1997; and others).

B. Water Quality

Regulation of the Colorado by a series of large dams has substantially increased stream salinity by two processes: the evaporation surface of the reservoirs and irrigation return flows (Pontius, 1997). Evaporative losses from the Colorado River reservoirs are especially high because of the arid climate of the region.

Salinity concentration is generally inversely proportional to flow rate, in that it decreases in periods of high flows and increases during periods of drought or otherwise induced low flows. Salinity levels have had significant domestic and

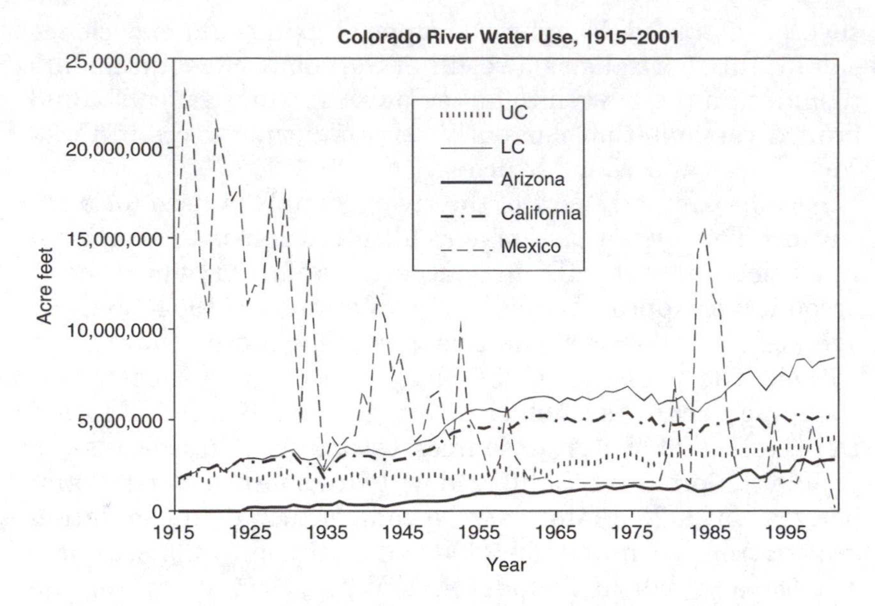

Figure 3 Trends in Colorado River use in the Upper and Lower Basins, 1915–2001. (Data from the U.S. Department of the Interior, Bureau of Reclamation.)

international impacts in the Colorado River basin. Because of the above-average precipitation in the Colorado watershed in the last several decades, high runoff and flood control releases have helped keep the river within standards set in the U.S.–Mexico treaty. In addition, Congress has taken a series of actions to control salinity. The salinity of the Colorado River water at its headwaters in the Rockies is about 50 mg of TDS (total dissolved solids) per liter. The stream salinity at the Mexican border doubled from 400 mg of TDS per liter in the early 1900s to 800 mg in the 1950s. About 50% of the salt in the river is from natural sources such as saline springs, erosion of saline geologic formations, and runoff, and the remainder comes from irrigation return flows (37%), reservoir evaporation and phreatophyte use (12%), and municipal and industrial effluent (1%) (Lane, 1998).

The 1944 international water treaty left important problems unresolved regarding the quality of water delivered by the United States to Mexico. The domestic impacts, such as pollution and low flow at source regions, resulted in a 1974 agreement in which the United States would assume costs for desalination of Colorado water before it enters Mexico. The agreement also has implications for water availability for the Colorado River delta during exceptionally dry periods.

In recent years, the stability and sustainability of the treaty apportionments have been challenged by three pressures (see Bennett and Herzog, 2000). The first is the demographic transformation underway in the border region. Since the passage of the North American Free Trade Agreement (NAFTA) in 1994, trade between Mexico and the United States has tripled to $261 billion, and with it the number of goods, vehicles, and services crossing the border has increased dramatically (INE, 2003). The second stress is environmental (habitat) considerations, and the third is drought.

Other water quality issues of recent concern along the Colorado include coliform contamination from inadequate waste treatment, limiting certain recreational activities, and perchlorate contamination that has leached into the water supply from an industrial point source near Las Vegas. Neither is directly related to drought, but they may have drought and water supply related implications.

III. THE CLIMATIC CONTEXT

The region encompassing the Colorado River basin poses special challenges for understanding and predicting weather and climate variability. Key factors include: complex terrain and correspondingly large topographic influences, multiple moisture sources and precipitation mechanisms, and large and variable water storage in the form of snowpack. Major variations in weather and climate extend across a broad temporal spectrum from daily through centennial timescales, with consequent effects on local and basin-wide hydrological budgets.

Longer term climate variations are also quite pronounced throughout the interior West and have major implications for the hydrology of the region. For example, the Bureau of Reclamation has estimated that water needs of the Lower Colorado

River Basin could not be met if the region were to experience a prolonged dry period such as occurred in the 1930s (el-Ashry and Gibbons, 1988). Paleoclimate evidence suggests that over the last two millennia several droughts occurred in this region that were of substantially greater severity and longer duration than any observed in the modern observational record, including the 1930s and the 1950s (Woodhouse, 2003).

For the western United States as a whole, approximately 50–70% of the annual precipitation falls in mountainous regions, mainly in the form of snow (Dracup, 1977). The Colorado is decidedly a snowmelt-driven system. Although much work on climatological drought has focused on precipitation amounts, for the Colorado, increases in temperature (which can be associated with drought as well as climate change) may be as important. Summer precipitation also provides an important moisture source for native ecosystems and dryland agriculture and reduces water needs for irrigated crops.

El Niño/Southern Oscillation (ENSO) events influence important aspects of the climate of the Colorado basin. ENSO events are the coupled anomalous oceanic warming (El Niño) and atmospheric response (Southern Oscillation) of the central and eastern tropical Pacific, known to affect climate worldwide. Its opposite phase, La Niña, is associated with anomalously cold ocean temperatures in the tropical Pacific. The general picture that arises from ENSO studies is that, in winter, El Niño conditions are associated with above-normal precipitation in the southwestern United States, including much of the Lower Basin, with a tendency toward below-normal precipitation in the Pacific Northwest. With La Niña conditions, the regional climate response is roughly the reverse, with below-normal precipitation more likely to occur in the southwest and above-normal precipitation expected in the Pacific Northwest. On average, in both El Niño and La Niña conditions, a nodal line in the wintertime response is located across central Colorado, indicating a tendency toward opposite-sign responses between the northern part of the Upper Basin and the Lower Basin. Decadal climate variability that affects the basin has been partly related to changes in the frequency and intensity of ENSO events and partly to a second mode of climate variability called the Pacific Decadal Oscillation, or PDO. In contrast to ENSO, PDO is more focused in the North Pacific extratropics. Several studies show statistically significant relationships between the PDO and streamflow in the western United States. They also identify significant multi-decadal shifts in moisture-controlled variables for the Upper Basin that were coincident with shifts in the PDO. The causes of the PDO are poorly understood. Clearly, if skillful forecasts of multi-year to decadal climate variability could be developed, they would have major applications for water resources planning and management in the basin.

At this time, confidence is very low in projecting long-term climate changes at regional scales, especially for precipitation. For temperatures, most climate change models are consistent in projecting wintertime warming over much of North America through this century (IPCC, 2001). Analyses of recent temperature trends have shown a tendency for warmer winters across the western United States since the mid-1960s (Livezey and Smith, 1999). Phenology studies, such as bloom dates for flowering lilac and honeysuckles, also indicate that spring blooms are occurring earlier than in the past through much of the West (Cayan et al., 2001). Even without changes in total precipitation, changes in the annual temperature cycle (e.g., a shortened cold season and lengthened warm season) could have significant implications for water resource use and management in the basin. Potential effects include changes in average annual snowpack (water storage) and evaporation, alterations in the magnitude and timing of the annual hydrological cycle (e.g., of peak flows), and additional water requirements to meet urban and agricultural needs.

The Powell Consortium (1995) study of the potential effects of severe sustained drought on the Colorado River system also brought out the importance of management flexibility in the face of extreme climate events. Existing institutional arrangements were found to protect traditional consumptive uses, but the nonconsumptive instream uses, such as hydropower and environmental requirements, were severely affected (Lord et al., 1995). Win–win solutions were possible over all water uses, but the study concluded that such possibilities were difficult to accomplish in practice. Given this background on climate and climate variations in the Colorado basin, we turn next to a discussion of four climate-sensitive water resources management problems within the basin.

IV. FOUR CLIMATE-SENSITIVE DECISION ENVIRONMENTS

A. International: The Border Region

Although international rivers have always been difficult to manage, the Colorado is especially interesting because of its enormously diverse and multiple overlapping jurisdictions, the strong contrast in legal and administrative styles of the two neighboring countries, and the exceptional degree of freedom and influence of the informal, nongovernmental sector in the United States (Varady et al., 2001).

In 1964, an international issue erupted when the Mexican government complained that deliveries of Colorado River water with salt concentrations of 2000 ppm were affecting crops and asserted that this was in violation of the 1944 Mexican Water Treaty . Salinity had become a major problem for Mexican agriculture in the Mexicali Valley after the 75,000-acre Wellton-Mohawk Irrigation District was developed in southern Arizona and the filling of Lake Powell had reduced flows in the river. After 10 years of negotiations, Mexico and the United States signed Minute No. 242 ("minute" in this context means an amendment to the 1944 treaty) in 1973, which established salinity standards for water delivered upstream of Morelos Dam (Mumme, 2000). The advantages included better relations between the United States and Mexico, with Mexico also waiving compensatory payments for historical damages.

Per Minute No. 242, the United States must deliver water to Mexico with an average annual salinity concentration no greater than 115 ppm +/– 30 ppm over the average annual salinity concentration of the river at Imperial Dam. Thus, an increase in salinity at Imperial Dam directly translates to an allowable increase in salinity of water delivered to Mexico and an increase in salinity of water flowing past Morelos Dam. Salinity is projected to increase at Imperial Dam to 980 mg/l by the year 2015 without additional controls (Bureau of Reclamation, 2002).

A parallel but more complex crisis is affecting much of the region's groundwater resources, which are largely outside the scope of the legal arrangements and beyond the control of most administrative agencies on both sides of the border. Although the states recognize the relationship between groundwater and surface water, their laws generally do not reflect this relationship. Groundwater use is poorly measured, but is generally acknowledged in many areas to exceed natural recharge. In times of low surface flow, water managers throughout the West tend to turn to groundwater as a backup supply. Because groundwater is frequently hydrologically connected to surface water, the generally unregulated use of groundwater frequently causes negative impacts on surface water users. Groundwater management issues are increasingly affecting the Colorado.

In December 2000, the two countries, acting through the International Boundary Waters Commission (IBWC), adopted Minute No. 306, recognizing a shared interest in the preservation of the riparian and estuarine ecology of the Colorado delta. Conflict over the delta has not fully developed in part because of wet episodes in the delta during the 1980s and 1990s. Despite extensive destruction, some recovery has been seen in the delta since 1981, when new flows coming from saline irrigation water or flood control operations were redirected, creating the Cienega de Santa Clara. This cienega has developed into an important habitat that is dependent on the continued irrigation return flows from the United States. Proposals by U.S. interests to operate the desalter at Yuma (built to treat Colorado River water to meet the standards in Minute 242, but never brought online) would increase water supply availability in the United States and meet U.S. obligations to Mexico. The relative roles of Mexico and the United States in resolving this issue are still evolving.

Recent efforts to deal with direct cross-border concerns include the Border Environmental Cooperation Commission (BECC) and the North American Development Bank (NAD Bank). The BECC and NAD Bank constitute a partial response to the water-related problems along the U.S.–Mexico border (Milich and Varady, 1999). The BECC offers a new kind of forum in which border residents are able to address problems they have in common. It is governed by a binational ten-member board of directors, which includes two members of the IBWC. Its charter explicitly emphasizes public participation. The BECC is charged with certifying proposed border infrastructure projects. BECC criteria include compliance with environmental requirements and maintenance of financial stability (Milich and Varady, 1999). Once a project is certified by the BECC it becomes eligible for financing by the NAD Bank. The BECC places regional proximity to the border ahead of national concerns. However, it is still too early to assess whether it can serve as a template for transboundary environmental institutions and whether there will be substantial implications for management of the Colorado River.

Under Minute 307 of the IBWC, the United States accepted Mexico's proposal for the two countries to cooperate in the fields of drought planning and sustainable use of the basin. However, in the United States, water rights and quantity management are generally the responsibility of states, not the federal government (Getches, 2003). Both surface water and groundwater are considered public resources subject to state law, with rights and permits to use water granted to individuals and water providers. Owners of water delivery and treatment infrastructure are typically not the states but local governments or private water companies and irrigation districts. A better understanding of the links between domestic concerns in both countries and international agreements is needed in order to construct a more complete picture of issues underlying cross-scale water-related disputes.

B. Arizona and California: Interstate Issues in the Lower Basin

The Colorado River is the principal source of water for irrigation and domestic use in Arizona, southern California, and southern Nevada. Accounting for the use and distribution of water from the Colorado River below Lees Ferry (lower Colorado River) is required by the U.S. Supreme Court Decree of 1964 in *Arizona v. California*. In addition to its other requirements, the Supreme Court decree dictates that the Secretary of the Interior (secretary) provides detailed and accurate records of diversion return flows and consumptive use of water diverted from the mainstream, "stated separately as to each diverter from the mainstream, each point of diversion, and each of the States of Arizona, California, and Nevada."

Arizona and California have a long history of battling over the Colorado. In 1964, after 11 years of legal battles, the U.S. Supreme Court, in *Arizona v. California*, confirmed the Upper and Lower Division apportionment of the Colorado. The court also held that Arizona's use of the Gila River and its tributaries would not reduce its entitlement of Colorado River water. A major concern for Arizonans has been protection of the state's allocation of Colorado River water from the other Lower Basin states (California and Nevada). Although *Arizona v. California* temporarily ended the battle for water supplies between the two states and quantified the rights to Colorado River water, California has been using approximately half a million acre-feet more water than its 4.4 maf allocation for many years. Along with concerns about the long-term reliability of Arizona's allocation, a conviction that Arizona needed to quickly utilize its full allocation developed during the 1980s and early 1990s, resulting in the creation of the Arizona Water Banking Authority (AWBA) in 1996.

The AWBA has four primary objectives: (1) to store water underground that can be recovered to ensure reliable municipal water deliveries during future shortages on the Colorado River or CAP (Central Arizona Project; discussed later in this chapter) system failures, (2) to support the management goals of the active management areas (AMAs; discussed later in this chapter), (3) to support Native American water rights settlements, and (4) to provide for interstate banking of Colorado River water to assist Nevada and California in meeting their water supply requirements while protecting Arizona's entitlement. The AWBA uses a combination of groundwater withdrawal fees, property taxes, and state general funds to purchase excess CAP water and contract with recharge facilities to store the water underground in central Arizona. The AWBA has been hailed as a major innovation in water management, and it has changed the tenor of inter-state negotiations substantially.

1. The Quantification Settlement Agreement (QSA)

Although the AWBA did help relieve some pressure among the Lower Basin states and provide a tool for responding to shortages during drought in Arizona, it did not resolve the basic problem of California's excess use of Colorado River water. In 2001, after years of complex inter-state discussions and a failed attempt by California to negotiate a multi-party intra-state agreement to address the overuse issue, Gail Norton, the Secretary of the Interior, required California to reduce its Colorado water use to its original apportionment. The Secretary's action served as a "focusing event" because it forced all of the parties back to the table to negotiate further.

The resulting agreements, signed October 10, 2003, between southern California water agencies, the State of California, and the federal government form the foundation of what is known as the California 4.4 Plan. Under a seven-state agreement to change the surplus criteria for managing the Colorado, California now has until 2017 to reduce its draw on the river from about 5.2 maf to its basic annual apportionment (4.4 maf) in the absence of surplus water. This "soft landing" is accomplished by renegotiating the interim surplus guidelines, which may exacerbate drought vulnerability by drawing down reservoirs farther than they would otherwise have been. The basic principle of the approach is that those who benefit from the interim surplus criteria (California) must also mitigate for the incremental harm to others (Arizona and Nevada) (Lochhead, 2003).

The Quantification Settlement Agreement (QSA) maps out how California will reduce its overreliance on the Colorado River while meeting the state's changing water needs. In particular, Colorado River water would shift from agricultural use (primarily within the Imperial Irrigation District and Coachella Valley Irrigation District, which hold the oldest priority water rights) to urban use (generally, Metropolitan Water District). In any event, even with initial hiccups (see Bureau of Reclamation, 2003), the negotiated solution of the California Plan and the interim surplus guidelines represents a remarkable achievement in good faith public interest negotiation-management (Lochhead, 2001, 2003). Such a solution is obviously preferable to litigation and competition between states and agencies, although it probably would not have happened without external forcing. It also illustrates the importance of water continuing to be a public resource rather than a private commodity (Lochhead, 2003). As this case illustrates, there is still flexibility in the system to accommodate changing needs and climatic conditions, although the level of effort required to develop agreements among the multiple affected parties is extremely high. Water use, efficiency, and transfers must be maximized locally before proceeding to the regional, inter-state, or inter-basin levels. Significant hurdles still must be overcome if inter-basin marketing is to become a reality.

C. Native American Water Rights

Thirty-four Indian reservations are located within the Colorado River basin, with the status of their water claims ranging from quantified in court, quantified through negotiated settlements, or still unquantified (Pontius, 1997). A number of tribes located outside the boundaries of the basin, such as the Mescalero Indian Reservation in New Mexico, have traditional or aboriginal interests in the basin as well. Each of these 57 reservations has very different interests, needs, and desires concerning the management of the Colorado River (Gelt, 1997; Pontius, 1997). The 1908 the *Winters v. United States Supreme Court* decision established the doctrine of Indian reserved water rights. The court held that such rights existed whether or not the tribes were using the water and dated to the time that the reservations were created. This decision was reaffirmed by *Arizona v. California* (1963), which awarded water rights to five Indian reservations in the Lower Basin. The court determined that an Indian tribe's quantified reserved right must be taken from and charged against the apportionment of water of the state where the tribe's reservation is located. Large outstanding Indian water rights claims in the Colorado River basin include Gila River (Arizona), 1,599,252 af; Hopi (Arizona), 140,406 af; Navajo (Arizona), 513,042 af; Tohono O'odham (Arizona), 650,000 af; and White Mt. Apache (Arizona), 179,847 af. The Gila River and Tohono O'odham settlements are included in a package that is currently (2004) being considered by Congress. If approved, the Gila River Settlement will be the largest in U.S. history, involving 643,000 af of water, multiple parties, and multiple side agreements.

The settlement of *Arizona v. California* had significant long-term implications for water management in the Colorado River basin. First, this case established the process for quantifying Winters' rights, potentially resulting in relatively large amounts of water for Indian tribes. Second, this case placed Indian water rights squarely within the framework of western water law, not only by quantifying the rights but also by holding that the Colorado River Indian tribes were included in Arizona's apportionment. This landmark decision means that Indian water rights were put in direct competition with other users within state allocations, increasing the pressure on surface water supplies, especially during drought. However, to the degree that Indian settlements result in the ability of other users to lease water from the tribes, these settlements will be a major source of water for municipal and industrial uses in the future.

D. Conjunctive Use and Management: Groundwater and Surface Water in Arizona

Conjunctive use is a term employed in multiple contexts. For the purpose of this chapter, the term is used to mean the integration of surface and groundwater supplies in order to maximize water supply availability. One mechanism for doing this is storing excess surface water in aquifers during times of ample supplies, with the expectation of recovery during dry years. This method of storage, though relatively inexpensive compared to construction of surface reservoirs, is dependent on the geology of the aquifers and the geography of water use patterns. Arizona has developed a number of institutional arrangements that facilitate artificial recharge and long-term water banking.

Most western states do not statutorily recognize artificial groundwater recharge as a beneficial use of water. However, in practice, artificial recharge is deemed to be of great benefit, because water can be stored relatively inexpensively with low evaporative losses, followed by recovery through the use of wells. Groundwater currently supports roughly half the total annual water demand in Arizona, with surface water, including diversions from the Colorado River, representing the other half (Jacobs and Holway, 2004). Before the completion of the Central Arizona Project (CAP), Arizona's use of Colorado River water was limited to diversions along the river itself, primarily for irrigation. Approximately 70% of the water use in the state is agricultural, although this percentage is expected to continue to decline over time, especially as cities grow in size. Arizona's population growth rate is among the highest in the nation; the population will be near 6 million in 2025, approximately three times that in 1980.

Throughout the West, groundwater is being pumped at rates that exceed the natural recharge rate. Arizona, California, Idaho, Nevada, and New Mexico have enacted comprehensive artificial groundwater recharge legislation to provide for growing needs. Groundwater levels have been dropping for decades, and recently states and utilities have begun recharge projects to replenish this diminishing resource. Artificial recharge is one way to offset these declines and manage the potential for subsidence, while responding to climatic variability in the surface water supply availability.

The CAP is designed to bring 1.415 maf of Arizona's 2.8 maf Colorado River allocation from Lake Havasu into central and southern Arizona. Deliveries to Phoenix began in 1985 and to the Tucson area in 1992. The CAP system is interconnected with the Salt River Project system in the Phoenix area, providing maximum flexibility for conjunctive management. However, the CAP has the lowest priority of the Lower Colorado allocations and must curtail its usage first in a shortage year. Concerns about the implications of the low priority of Arizona's Colorado River water and the overallocation of its supplies have driven a number of innovations in the context of inter-state negotiations, such as the AWBA and discussions of resource reliability in the interim surplus guidelines.

Within Arizona, municipal CAP deliveries have higher priority than agricultural deliveries, so agriculture will be affected first if there is a shortage to the CAP. The likelihood of curtailment of deliveries to municipal interests due to shortfalls on the Colorado in the next 30 years is considered by CAP to be very limited, primarily because the Upper Basin states (Colorado, Utah, New Mexico, and Wyoming) have not fully developed use of their allocations. However, recent severe drought conditions affecting the Colorado and the Salt River system simultaneously, the interim surplus agreement with California, which results in lower mainstem reservoir levels, and predictions of a possible decades-long drought have raised the level of concern about curtailment in both the near and long term.

The 1980 Groundwater Management Act (GMA) changed the institutional arrangements for managing groundwater in Arizona in several dramatic ways. The focus of the GMA provisions is within active management areas (AMAs), which are portions of the state where the majority of the population and groundwater overdraft are concentrated. The management goal for all of the AMAs focuses on developing a sustainable water supply. In the case of the major metropolitan areas, the goal is "safe yield." The AMAs include more than 80% of Arizona's population, more than 50% of total water use in the state, and 70% of the state's groundwater overdraft, but only 23% of the land area. The GMA uses a primarily regulatory approach to managing groundwater supplies. The program includes mandatory reductions in demand for all sectors through conservation and a required transition to renewable supplies.

A key component in encouraging the use of Colorado River water is the Assured Water Supply (AWS) program, which requires that all new subdivisions in AMAs demonstrate a 100-year AWS based primarily on renewable water supplies before the subdivision is approved. This long-term planning horizon has proved challenging in the face of high growth rates and variable surface water supplies but has caused substantial investment in the use of renewable supplies.

Implementation of the AWS rules in 1995 would likely not have been politically feasible in Arizona without the provision of a convenient mechanism for most residential developers, particularly those without ready access to renewable supplies, to continue building. The Central Arizona Groundwater Replenishment District (CAGRD), by committing to replenish groundwater used by its members, provided a mechanism to meet the requirement to use renewable supplies and therefore is partially responsible for the ability to adopt relatively stringent AWS rules. The popularity of the CAGRD has exceeded all expectations, leaving the agency working hard to ensure a reliable long-term water supply for all its customers and to find sufficient recharge capacity in each of the AMAs.

With the creation of the AWBA in 1996, the development of incentive pricing programs for agriculture and recharge, and the AWS rules in place, Arizona is now fully utilizing its Colorado River allocation. Annual utilization patterns are strongly affected by surface water supply conditions within the state, agricultural demand, and availability of Colorado River water, but the full allocation has been diverted for the past several years. Although importing surface water has significant benefits from the perspective of relieving pressure on diminishing groundwater supplies and ensuring a long-term water supply, it has also increased the vulnerability of the state to climate variability. The strong focus on artificial recharge is clearly accepted as a tool to offset future drought-related shortages.

V. OPPORTUNITIES FOR TECHNOLOGICAL INTERVENTIONS AND CLIMATE SCIENCE APPLICATIONS

The three major elements of western states' water future are (1) conservation and demand management, (2) municipal–agricultural cooperation, and (3) supply integration. Conservation and demand management approaches range from technology interventions for specific problems to regional water basin planning, including mandatory, voluntary, and incentive-based approaches (Luecke et al., 2003). The innovations described above (the AWBA/AWS/CAGRD, BECC, the QSA settlement) are based in water management planning and provide institutional mechanisms to reduce vulnerability to drought, potentially limiting the economic impact of shortfalls in Colorado River deliveries.

Gleick (2003) describes the rise of "soft path" approaches that complement physical infrastructure with lower cost community-scale systems, decentralized and open decision making, water markets where actually needed, equitable pricing, application of efficient technology, and environmental protection. Given the lack of sites left for new dams on the Colorado and the economic and environmental costs associated with dams, soft path approaches are widely viewed as viable alternatives to supply enhancement. The Council of State Governments (2003) in a recent report identified several such "soft path" mechanisms being employed to different degrees to combat overuse. These include pricing structures to promote water use efficiency; measurement of water usage; audits of commercial, domestic, and industrial uses; water reuse and recycling; management of water system pressure; retrofitting and replacement of water fixtures; promotion of water-efficient appliances; improving infrastructure quality; conservation; and conservation education. As Gleick et al. (2002) showed, increased economic growth does not always require increased water development, but trends in management still reflect this traditional belief. In addition, there is mounting evidence that in the fastest-growing regions of the West, increased storage (in the absence of a full investigation of water supply reliability during drought) simply encourages increased development during times of plenty as opposed to acting as a buffer for drought (Luecke et al., 2003; Pulwarty, 2003).

A. Opportunities for Application of Climatic Information

Managers have traditionally relied on the historical record in order to plan for the future, inferring the probability that shortages and floods might occur given their frequency of occurrence in the recent past. Problems are further compounded by lack of agreement on definitions and concepts, such as "extraordinary drought" and "optimal utilization." Water managers in the basin have developed tools for dealing with risk and uncertainty, mostly derived from relatively short climatic records (<100 years). As is clear from numerous paleoclimatic records and sources, climate has never been "stable" for long periods, even if we have created statistical artifacts such as climate averages and event recurrence estimations based on short records. For example, in most parts of the Colorado basin, reliable flow measurements for major streams have been recorded only over the last 50–100 years and precipitation measurements over the last 20–60 years. Water managers often lack even basic data on water quantity and quality, the nature of climate variations, and their impacts on water users and uses, and thus have little basis for designing effective management programs (Jacobs and Pulwarty, 2004). More specific forecasts are needed for different regions and sectors to assist water managers in proactive planning. Climate forecasts are now available on biweekly, monthly, and seasonal to interannual scales and are improving in skill over time. Demand forecasts are equally important and need to be undertaken for 5-, 10-, and 20-year horizons. Given recent advancements in understanding climate variability and change, it is clear that such projections must be made in the context of the greater than 10 years timescales of climate variations that exist in the Colorado system. Water managers have differing needs for scientific information relative to the scale of management, the type of decision being made, and the nature of the decision (e.g., long-term investments vs. short-term operational decisions). In the case of a large watershed such as the Colorado, these factors cross several time and space scales. However, on the climate side, substantial work is still needed to increase predictive capability (and appropriate applications) at the regional scale, especially where there is substantial topographic variability. Preliminary approaches have included both demonstration experiments in the use of climate information and assessment of impediments to the flow of information in practical settings (Georgakakos, 2002; Pulwarty and Melis, 2001). At the level of small watersheds it becomes extremely important not to oversell the precision of forecasts at the expense of being clear about their accuracy. Thus scaling up from local data is as important as scaling down from globally forced regional models.

VI. PRESENT CONDITIONS ON THE COLORADO: SITUATION "NORMAL" = SITUATION "CRITICAL"

The U.S. Geological Survey has confirmed that the 1999–2004 period is the driest in the almost 100 years of recorded streamflow history of the Colorado River. Lake Powell was full (24 maf) in 1999. With below-average snowfall in March 2004, the estimated Colorado River inflow into Lake Powell for 2004 is at 50% of average (Bureau of Reclamation, 2004). In April 2004 the lake was at 10.2 maf (about 42% capacity), a level not seen since 1971, when the lake was still being filled; Lake Mead was at 59% capacity. Given current demand, an average inflow year will increase system storage by only about 3% per year. Even if Lake Powell does not empty, Bureau of Reclamation officials estimate that it will take a minimum of 13 years to refill with average precipitation (Keys, 2003).

The delivery obligation to the Lower Basin is fixed. If Lake Powell were to dry up, then cuts might have to be made to western slope transfers to the Front Range. Colorado, for instance, uses about 2.5 maf of water a year, of which about 0.5 maf is transferred to the growing Front Range through inter-basin diversions from the Colorado to the South Platte and Arkansas Rivers. Colorado could "theoretically be responsible for 51.75% of any shortfall" (J. Lochhead, *Denver Post*,

April 2, 2004). At the present rate of runoff (over the past 5 years), hydrologists estimate that Lake Powell could be dry by 2007. For the first time since Hoover Dam was built in the 1930s, the states that depend on the Colorado are preparing for the possibility of shortages. Without an alternative plan, *Arizona v. California* 1963 could trigger measures that would significantly reduce Arizona's water from the Colorado. The spatial extent and persistence of the present drought may be such that the Upper Basin may not be able to produce the Lower Basin requirements without cuts of its own, as required by the 1922 compact.

The massive plumbing network built to serve the exploding population of western states has removed much of the buffer that was available in earlier decades. The grace period would be about 2 to 3 years at the current inflow rate before the Secretary of the Interior declares a shortage. As noted by one legal scholar with a long history of involvement in western water resource issues, "if we think that we have more Colorado River water to develop then we had better think again" (David Getches, *Denver Post*, April 2, 2004). The complex set of compacts, congressional acts, and case law that governs the river has never been tested by the kind of extended drought that Lake Powell, Lake Mead, and the other reservoirs were designed to guard against. According to Bennett Raley (Assistant Secretary, Department of the Interior, *Denver Post*, April 4, 2004), "The crises we face will be (in) normal years and they will be about meeting existing demands ... the fight is no longer about water decades into the future."

VII. CONCLUSION

An overarching question throughout this chapter has been "Are assumptions about planning in the Colorado basin borne out by the climate record and by projections of change?" The answer would have to be "No." Representatives from the seven basin states have, at least recently, recognized that there is a finite time for response (see *New York Times*, May 2, 2004; Department of the Interior, 2003).

Extreme events are the chief drivers of water resources system adjustments to environmental and social change. How well water systems handle the extreme tails of current or altered climate distributions is likely to be an overriding concern as systems become more constrained. The behavioral/institutional problem is that decision makers at different levels of governance, researchers, and resource managers have difficulty anticipating how complex systems will respond to environmental stresses and may not have the flexibility to respond when they do understand key areas of vulnerability. Historically, water banking and inter-basin transfers have been used to mitigate the effects of short-term drought in the Colorado basin. The lessons and impacts of these adjustment strategies are still being gathered. However, the maintenance of supply during periods of severe long-term droughts of 5–10 years to multiple decades (the timescales of development, project implementation, and ecosystem management efforts), known to have occurred in the West over the past 1000 years, is as yet untested but may be so in the very near future.

A major stumbling block for comprehensive water management is the adversarial relationship that usually develops between upstream and downstream users of water. Even in areas where integrated approaches were adopted, cooperation remains mainly crisis driven, inhibiting iterative, long-term collaboration and learning. As noted above, although opportunities for "win–win" situations and rule changes exist, such changes are extremely difficult to implement. The experience of development of the Colorado in the face of environmental uncertainty clearly illustrates that impacts and interventions can reverberate through the systems in ways that can only be partially traced and predicted. The failure to consider the system implications of policy and environmental changes can mean that policies actually have long-term counterproductive effects that can decrease rather than increase system performances for economy and environment (Innes and Booher, 2003). A similar observation from the disasters research community is that short-term adjustments to extreme events can increase and have increased longer term vulnerability (Comfort et al., 1999). Recommendations based on technocratic interventions or economic efficiency must be conditioned within the practical ranges of choices actually available to decision makers, cognizant of legal and cultural constraints. Developing a good understanding of the climate-sensitive policy decision environments is one of the most difficult challenges to developing effective interventions of science and technology and of cooperative strategies.

Varady and Morehouse (2003) point out that what is distinctive about the Colorado is that the inclusion of stakeholders in water management policy has become the norm. However, as the noted water resource economist Charles Howe (personal communication) observes, regardless of how robust civil-society institutions may be, severe drought (or flooding) usually exposes underlying institutional barriers to effective cooperation. Major increases in natural resource demands on either side of a border (from population, industry, or commerce, or some combination of these factors) often confront contradictions embedded in notions of sovereignty, local control, or other such institutional arrangements. Thus for large river basins the goal should not be to emphasize some particular scale of analysis (e.g., local, regional) or approach such as decentralization or centralization, at the expense of other or competing problem definitions. The goal, instead, should be to uncover what is needed at each of these scales and to address impediments and opportunities to the flow of information and innovations between the decision-making nodes.

Most decision makers engaged in cooperative strategies addressing water scarcity have repeatedly stated the need for integrated management of existing supplies and infrastructure (see Frederick et al., 1996). To provide the best available information for decision support for managing crises or at least providing acceptable outcomes in such situations, we repeat two recommendations made in numerous fora throughout the West (see Pontius, 1997; Szekely, 1992, 1993; and others). The first is the need for a centralized and integrated data center for the Colorado River basin. Such a center should be established to (1) collect and provide a comprehensive, reliable, scientific and economic, demographic, and cultural database, including use patterns, economic activity and productivity by sector, environmental health statistics, and other critical data; (2) produce syntheses of best available knowledge about interactions between climate/hydrology and water resource supply and demand along the border; and (3) develop an evaluation of socioeconomic data and projections of socioeconomic and demographic changes over the next 25 years in terms of potential impact on water supply and demand.

As Wilhite repeatedly warns, reactive mechanisms such as drought relief do little if anything to reduce the vulnerability of the affected area to future drought (see, e.g., Wilhite et al., 2000). Thus, if the outcomes of the above recommendation are to prove useful, then a receptive decision environment coordinated by a council with oversight of the entire basin would be advantageous. This second recommendation—the formation of a Colorado River Basin Coordinating Council—has been suggested before (Getches, 1997; Weatherford and Brown, 1986). Such a council could be self-supporting from hydropower revenues and water delivery charges, among other things (McDonnell and Driver, 1996). It could emerge only from an evolutionary process (Pontius, 1997). Crisis can create political consent or act as a catalyst for change even while crisis management itself is usually ineffective in the long term. The impetus of the recent drought has not only focused attention on the disturbing water resource trends at different scales of use, but also illuminated the shortcomings of the existing water management frameworks, which were effectively designed but for a different era.

REFERENCES

Bennett V, L Herzog. U.S.–Mexico borderland water conflicts and institutional change: A commentary. *Natural Resources Journal* 40:973–989, 2000.

Booker J, R Young. Modeling intrastate and interstate markets for Colorado River water resources. *Journal of Environmental and Economic Management* 26:66–87, 1994.

Bureau of Reclamation. Colorado River System Consumptive Uses and Losses Reports 1995–2000. Available from U.S. Department of the Interior, Bureau of Reclamation. *http://www.usbr.gov/uc/library/envdocs/reports/crs/crsul.html*, 2001.

Bureau of Reclamation. Colorado River Basin Salinity Control Program. U.S. Department of the Interior, Bureau of Reclamation. Annual Accountability Report 2002, 2002.

Bureau of Reclamation. Interior Sends California Agencies Revised Water Orders, Letter on Progress in Reviewing QSA, and Initiates de novo Part 417 Proceedings. U.S. Department of the Interior, Bureau of Reclamation. News Release 3 April, 2003.

Bureau of Reclamation. Reclamation Drought Update. Office of the Commissioner. U.S. Department of the Interior, Bureau of Reclamation. News Release 23 April, 2004.

Cayan D, S Kammerdiener, M Dettinger, J Caprio, D Peterson. Changes in the onset of spring in the western United States. *Bulletin of the American Meteorological Society* 82:399–415, 2001.

Comfort L, B Wisner, S Cutter, R Pulwarty, K Hewitt, A Oliver-Smith, J Wiener, M Fordham, W Peacock, F. Krimgold. Reframing disaster policy: The global evolution of vulnerable communities. *Environmental Hazards* 1:39–44, 1999.

Council of State Governments. Water Wars: Trends Alerts. Lexington, KY, 2003.

Department of the Interior. Water 2025: Preventing Crises and Conflict in the West. *http://www.usbr.gov/centennial/*, 2003.

Dracup J. Impact on the Colorado River Basin and Southwest water supply. In: *National Research Council, Climate Change and Water Supply* (p. 121). Washington, D.C.: National Academy Press, 1977.

el-Ashry M, D Gibbons. *Water and Arid Lands of the Western United States.* World Resources Institute. Cambridge University Press, 1988.

Frederick K, T Vandenberg, J Hanson. Economic Values of Freshwater in the United States. Resources for the Future RFF Press, Washington, D.C., 1996.

Gelt J. Sharing Colorado River Water: History, Public Policy and the Colorado River Compact. *Arroyo*, 10, No. 1. Water Resources Research Center. University of Arizona, Tucson, 1997.

Georgakakos K. Climate forecasts and water resources management: A fertile field for hydroinformatics. Keynote paper. International Conference on Hydroinformatics. Cardiff, U.K., 1–5 July 2002.

Getches D. Colorado River governance: Shared federal authority as an incentive to create a new institution. *University of Colorado Law Review* 68(3), 1997.

Getches D. Impacts in Mexico of Colorado River management in the United States: A history of neglect, a future of uncertainty. In: H Diaz and B Morehouse (Eds.), *Climate and Water: Transboundary Challenges in the Americas* (pp. 163–192). Netherlands: Kluwer, 2003.

Gleick P. Global freshwater resources: Soft-path solutions for the 21st century. *Science* 302:1524–1528, 2003.

Gleick P, G Wolff, E Chalecki, R Reyes. *The New Economy of Water.* Oakland, CA: Pacific Institute, 2002.

INE. Instituto Nacional de Estadística, Geografía e Informática. Industria Maquiladora de Exportación. Estadísticas Económicas. Abril 2003. *http://www.inegi.gob.mx*, International Boundary and Water Commission, 2003.

Innes J, D Booher. Collaborative policymaking: Governance through dialogue. In: M Hajer, H Wagenaar (Eds), *Deliberative Policy Analysis: Understanding Governance in the Network Society* (Chapter 1). London: Cambridge University Press, 2003.

IPCC. *Climate Change 2001: The Scientific Basis. United Nations Intergovernmental Panel on Climate Change Third Assessment Report*. Port Chester, N.Y., Cambridge University Press, 2001.

Jacobs K, J Holway. Managing for sustainability in an arid climate: Lessons learned from 20 years of groundwater management in Arizona, USA. *Hydrogeology Journal* 12, 52–65, 2004.

Jacobs K, R Pulwarty. Water resources management: Science, planning and decisionmaking. In: R Lawford, D Fort, H Hartmann, S Eden (Eds), *Water: Science, Policy and Management*. Washington, D.C.: American Geophysical Union Press, 2004.

Keys J. Drought or Opportunity. Remarks of the Commissioner. U.S. Department of the Interior, Bureau of Reclamation. Colorado River Water Users Association, Las Vegas, NV, December 12, 2003.

Lane WL. Statistical Analysis of the 1995 Lower Colorado River Accounting System, An Assessment of Current Procedures with Recommended Improvements (available from the Bureau of Reclamation, Boulder Canyon Operations Office, Boulder City, NV), February 1998.

Livezey R, T Smith. Covariability of aspects of North American climate with global sea surface temperatures on interannual to interdecadal timescales. *Journal of Climate* 12:289–302, 1999.

Lochhead J. An Upper Basin perspective on California's claims to water from the Colorado River. Part 1: The law of the river. *University of Denver Water Law Review* 4:291–310, 2001.

Lochhead J. An Upper Basin perspective on California's claims to water from the Colorado River. Part II: The development, implementation and collapse of California's plan to live within its basic apportionment. *University of Denver Water Law Review* 6:318–400, 2003.

Lord W, J Booker, D Getches, B Harding, D Kenney, R Young. Managing the Colorado River in a severe sustained drought: An evaluation of the institutional options. *Water Resources Bulletin* 31:939–944 (in the Powell Consortium study), 1995.

Luecke D, J Morriss, L Rozaklis, R Weaver. What the Current Drought Means for the Future of Water Management in Colorado. Land and Water Fund of the Rockies. Denver, CO, 2003.

McDonnell L, B Driver. Rethinking Colorado River Governance. The Colorado River Workshop (pp. 181–222). Grand Canyon Trust, Phoenix, AZ, 1996.

Milich L, R Varady. Openness, sustainability, public participation: New designs for transboundary river institutions. *Journal of Environmental Development* 8:258–306, 1999.

Morrison J, S Postel, P Gleick. The Sustainable Use of Water in the Lower Colorado River Basin. Pacific Institute for Studies in Development Environment and Security, Oakland, CA, 1996.

Mumme S. Minute 242 and beyond: Challenges and opportunities for managing transboundary groundwater on the Mexico–U.S. border. *Natural Resources Journal* 40:341, 2000.

Pontius D. Colorado River Basin Study. Final Report to the Western Water Policy Review Advisory Commission. National Technical Information Service Springfield, VA, 1997.

Powell Consortium. Severe sustained drought: Managing the Colorado River in times of shortage. *Water Resources Bulletin* 31, (full issue), 1995.

Pulwarty R. Climate and water in the West: Science, information and decisionmaking. *Water Resources Update* 124:4–12, 2003.

Pulwarty R, T Melis. Climate extremes and adaptive management on the Colorado River. *Journal of Environmental Management* 63:307–324, 2001.

Stegner W. *Beyond the Hundredth Meridian.* Boston: Houghton Mifflin, 1954.

Stockton C, W Boggess. Geohydrological Implications of Climate Change on Water Resource Development. U.S. Army Coastal Engineering Research Center. Fort Belvoir, VA, 1979.

Székely A. Establishing a region for ecological cooperation in North America. *Natural Resources Journal* 32:563–622, 1992.

Székely A. How to accommodate an uncertain future into institutional responsiveness and planning: The case of Mexico and the United States. *Natural Resources Journal* 33:397, 1993.

Varady R, B Morehouse. Moving borders from the periphery to the center: River basins, political boundaries, and water management policy. In: R Lawford, D Fort, H Hartmann, S Eden (Eds), *2004: Water: Science, Policy and Management* (pp. 143–160). Washington, D.C.: American Geophysical Union Press, 2003.

Varady R, A Kaus, R Merideth, J Campoy, K Hankins. The rise of stakeholder influence in environmental decision making in the U.S.–Mexico border region: Livelihood, conservation, and water use in the Lower Colorado River and Delta. Paper presented at "Scientists and Decision Makers: Acting Together for Sustainable Management of Our River Systems." Agence de l'Eau Rhone-Mediterranee-Corse, Lyon, France, 6 June 2001.

Weatherford G, F Brown (Eds.). *New Courses for the Colorado River.* Albuquerque, NM: University of New Mexico Press, 1986.

Wilhite D, MJ Hayes, C Knutson, KH Smith. Planning for drought: Moving from crisis to risk management. *Journal of the American Water Resources Association* 36:697–710, 2000.

Woodhouse C. A 431-year reconstruction of Western Colorado snowpack. *Journal of Climate* 16:1551–1561, 2003.

11

Drought Risk Management in Canada–U.S. Transboundary Watersheds: Now and in the Future

GRACE KOSHIDA, MARIANNE ALDEN,
STEWART J. COHEN, ROBERT A. HALLIDAY,
LINDA D. MORTSCH, VIRGINIA WITTROCK,
AND ABDEL R. MAAROUF

CONTENTS

I. INTRODUCTION

The water resources of Canada and the United States have been heavily modified and intensively managed to serve a variety of human needs. Because most human activities and ecosystem health depend on reliable, adequate water supplies, droughts present a serious threat to water management. Large-area droughts heavily impact a wide range of water-sensitive sectors, including agriculture, energy production, industry, municipalities, tourism and recreation, and aquatic ecosystems. They often stress local and regional water supplies by reducing streamflows, lowering lake and reservoir levels, and diminishing groundwater supplies (Bonsal et al., 2004).

Although droughts occur throughout Canada and the United States, some of the most severe and widespread droughts take place on the North American Great Plains, a semiarid to subhumid area that experiences highly variable weather. Severe droughts and heat waves occurred in this region in 1988, 1990, and 2001–2002 (Bonsal et al., 2003; Etkin, 1997). By contrast, droughts in eastern North America are usually brief, cover a smaller area, and are less frequent and severe (Koshida et al., 1999).

In Canada and the United States, laws and institutions exist that govern allocation of water among competing uses and define the rights and obligations of individuals and governments with respect to particular water resources. Disputes in transboundary watersheds along the Canada–U.S. border are handled under the Boundary Waters Treaty of 1909. The International Joint Commission (IJC), responsible for implementing the treaty, is a quasi-judicial body comprising six members, three appointed by each country. The IJC has two main functions: (1) to review applications for changes in flows and levels in boundary waters and issue Orders of Approval, and (2) to provide nonbinding advice on any question brought by governments (IJC, 1990).

Drought risk is a combination of a region's exposure to the drought hazard and its vulnerability to extended periods of water shortage. With dozens of watersheds currently being shared between Canada and the United States, droughts can exacerbate existing international water conflicts between the two countries.

II. FUTURE THREATS AND IMPACTS

Vulnerability to drought is influenced by a wide range of factors such as natural resource management policies, water use trends, land use, urbanization, and government policies. The current water management infrastructure has allowed the citizens of both Canada and the United States to make productive use of water and reduce the adverse impacts of droughts and floods.

However, population growth and continuing urbanization will affect the long-term availability of water supplies, future levels of demand, and the longevity and robustness of water supply and distribution infrastructure. Any changes in the quantity, quality, or timing of water supplies could also pose serious threats to human health and well-being.

Lack of clean water and sanitation services increases the risk of infectious diseases such as diarrhea, *Cryptosporidium*, *Giardia*, typhoid fever, hepatitis A, and cholera. Wastewater could mix with water supplies, allowing bacteria, viruses, and parasites to leak into drinking and cooking water and food. A severe outbreak of a deadly strain of *E. coli* in Walkerton, Ontario, in spring 2000, causing several deaths and thousands of disease cases, was triggered by an extended period of dry weather followed by heavy rains washing cattle fecal matter into well water (Chiotti et al., 2002).

Droughts can exacerbate conditions by causing disease-carrying rodents, such as deer mice that transmit Hantavirus, to leave their natural habitats and move into populated areas searching for food. Spring and early summer droughts were also found to precede severe outbreaks of West Nile virus (Epstein and Defilippo, 2001). During droughts, many sources of water dry up and the remaining sources of water become gathering spots for birds, which become easy targets for mosquitoes.

Forests become more susceptible to fires during droughts. Widespread bush fires in combination with low mixing heights and light winds can result in severe air pollution episodes, causing acute respiratory illness. Direct exposure to fires can lead to burning injuries or death.

Climate and water are closely linked because the global energy balance drives the distribution of moisture through precipitation and evaporation. Available evidence suggests that future climate change may lead to substantial changes in mean annual streamflows and seasonal distributions of flows. Climate change is expected to alter regional hydrologic processes and thus modify the quantity, quality, and timing of water resources. But the greatest impacts of climate change will likely be caused by an increase in the frequency and intensity of extreme events, such as droughts and floods (Cohen et al., 2001; Koshida et al., 1997).

III. CASE STUDIES

How has drought risk been managed in Canada–U.S. transboundary watersheds? The following three regional case studies illustrate past water management successes, controversies, and failures. Climate change impacts on water resources in Canada and the United States are expected to be significant. The question for water resource managers is whether the impacts of climate change will be large enough and will occur rapidly enough to require measures to be taken to adapt to their effects.

The case studies will use specific climate change scenarios and illustrate how principles of adaptive management can be incorporated to reduce a region's vulnerability to future climate extremes, including drought.

The scenario-generating technique most commonly used in climate change impact assessments is based on general circulation models (GCMs). In the scenario-generating process, the 30-year simulation period from 1961 to 1990 is used as the reference climate from which "change fields" for future periods are calculated. Most scenarios are calculated for the years 2020, 2050, and 2080. Three of the most popular GCMs are CGCM2 (Canada), CSIRO (Australia), and HADCM3 (United Kingdom) (Bruce et al., 2003).

A. Okanagan Basin

The Okanagan basin is located in the southern interior of British Columbia, Canada, situated around Okanagan Lake (Figure 1). The surface area of the basin is 8200 km^2 (Cohen and Kulkarni, 2001). The Okanagan has a dry continental climate, because the valley sits in the rain shadow of the Coast and Cascade Mountain ranges. The semiarid climate receives approximately 30 cm of precipitation each year, of which 85% is lost through evapotranspiration from local lakes. The

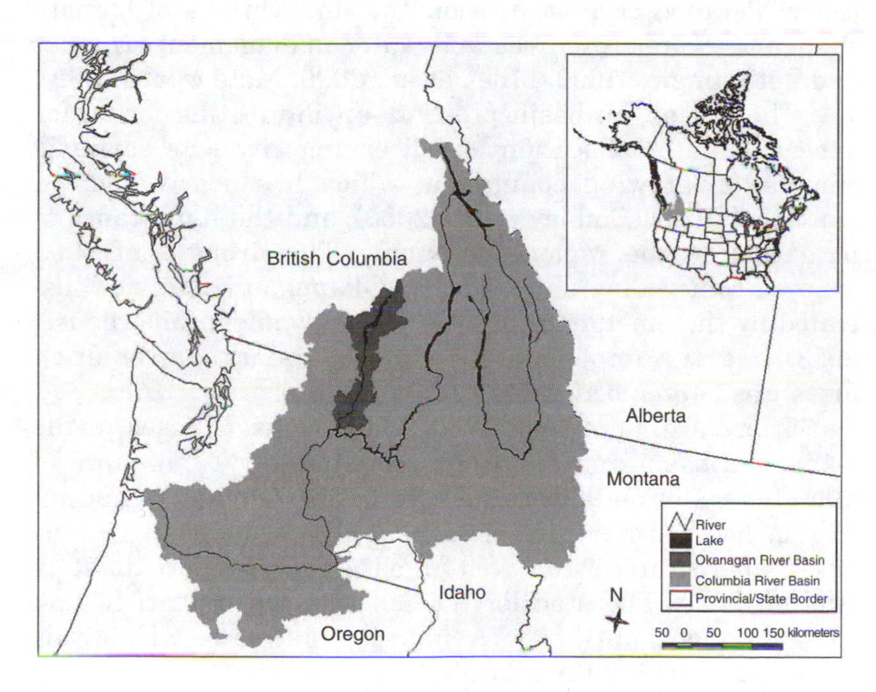

Figure 1 Okanagan and Columbia River basins.

hydrology of the basin is largely snow dominated, with much of the water that enters the lakes and Okanagan River originating from high-elevation regions (Cohen and Neale, 2003). The quality of water from these is generally good, but outbreaks of both *Giardia* and *Cryptosporidium* have occurred in valley communities (Cohen and Kulkarni, 2001).

The Okanagan region has become the most prominent location for 90% of soft fruit orchards and 95% of vineyards in British Columbia. The arid Okanagan summers are beneficial for fruit development but provide insufficient moisture, so irrigation is steadily used to support the growing crops. The region's extensive natural resources also contribute to its thriving tourism industry (Cohen and Kulkarni, 2001).

The Okanagan River flows south from British Columbia into Washington State, where it eventually meets the main stem of the Columbia River. The Columbia has been the subject of detailed case studies on the implications of climate change for water resources and water management (Hamlet and Lettenmaier, 1999; Miles et al., 2000; Mote et al., 1999).

The Okanagan basin presents an interesting forum for exploring water allocation and licensing given its semiarid climate; its growing population, which has nearly doubled since the 1970s (Embley et al., 2001); and the importance of irrigation to the regional economy. The drought of 2003 exposed some vulnerability in the Okanagan basin, as illustrated by the emergence of local water conflicts (Moorhouse, 2003) and the implementation of emergency conservation measures (*Watershed News*, 2003a, 2003b).

There are more than 4000 active water licenses in the Okanagan basin, representing approximately 1 billion m^3 of allocated water on 980 streams for both consumptive and instream uses. Around 45% is allocated for consumptive purposes, where water is removed from the source. Approximately two-thirds are allocated for the purposes of "irrigation" and "irrigation local authority." A majority of streams within this basin are already fully allocated.

Since 1997, a number of studies have been initiated on climate change, climate impacts, and adaptation within the Okanagan region (Cohen and Kulkarni, 2001; Cohen and

Neale, 2003; Cohen et al., 2000; Cohen et al., 2004; Merritt and Alila, 2004; Neilsen et al., 2001; Shepherd, 2004. The climate change scenario for the 2050s that appears to be emerging is as follows:

- A warming, relative to the 1961–90 baseline, of 1.5–4°C in winter with precipitation increases on the order of 5–25%; for summer, a warming of roughly 2–4°C and precipitation changes ranging from almost no change to a 35% decrease
- An earlier spring freshet of around 4 weeks, with reductions of annual and freshet flow volumes; for example, annual volumes for the Ellis reservoir near Penticton decline 20–35%
- An increase in crop water demand, due to warmer growing conditions, of 20–35% for the region as a whole; estimated increases of 20–40% for Oliver, near Osoyoos (D. Nielsen, personal communication)

This combination of a longer and warmer growing season, reduced water supply, and increased crop water demand represents a new average state for Okanagan water resources for the 2050s, and this suggests an increase in the frequency and severity of dry years with conditions likely to be considered as drought. This scenario does not assume any particular adaptation strategy, nor does it assume any changes in management practices in agriculture or among other regional bodies (municipalities, water agencies, fisheries interests, etc.). It does establish a "what if" context for consideration of possible options for adaptation.

A number of demand-side and supply-side options can be considered, including additional withdrawals directly from Okanagan Lake and the Okanagan River to augment withdrawals from the tributary streams and groundwater. Costs vary widely; from CAN$500 to CAN$3400 per acre-foot, and no single option would appear to be sufficient (see Cohen and Neale, 2003).

Previous discussions with regional stakeholders (see Cohen and Kulkarni, 2001) revealed no clear preference. Increased storage in upstream areas, buying back some existing water licenses, metering, and public efforts to reduce demand for water (e.g., through xeriscaping residential areas) are ideas that appear to have some support. There has been some recent experience with instituting metering in the city of Kelowna and in the Southeast Kelowna Irrigation District (Shepherd, 2004). Plans are underway for a new round of dialogue exercises with regional interests to consider how an adaptation portfolio might be developed and implemented (Cohen and Neale, 2003).

It is also conceivable that changes to operating rules could become part of an adaptation strategy to address the climate change scenario being considered here. Flow near the transboundary border is controlled by the Zosel Dam, built on the U.S. side just south of Osoyoos Lake and regulated by the IJC's Osoyoos Lake Board of Control. The Penticton Lake Dam controls outflows from Okanagan Lake. And several key reservoirs (such as Ellis reservoir) provide storage in the upstream areas.

This case represents an important opportunity to explore climate change adaptation in a proactive manner and in the context of ongoing planning processes that are a normal part of regional and local governance.

B. Poplar and Red Basins

The Poplar River (Figure 2) rises in Saskatchewan and flows southward, joining the Missouri River at Poplar, Montana. Roughly one-third of the basin is in Canada. The entire surface area of the basin is only 8620 km^2. Because annual evapotranspiration usually exceeds precipitation, the climate is considered semiarid. The mean annual flow of the Poplar, where it joins the Missouri, is 3.8 m^3/s, about three-quarters of that occurring as spring snowmelt runoff (IJC, 1978). This is equivalent to only 13 mm on the surface area of the basin. Significant variations in flow occur both within years and between years. At the international boundary, the natural flow of this river ranges from zero in the summer and autumn months to the record high monthly maximum of 19.9 m^3/s in April 1952. Since the 1990s, the maximum streamflow has

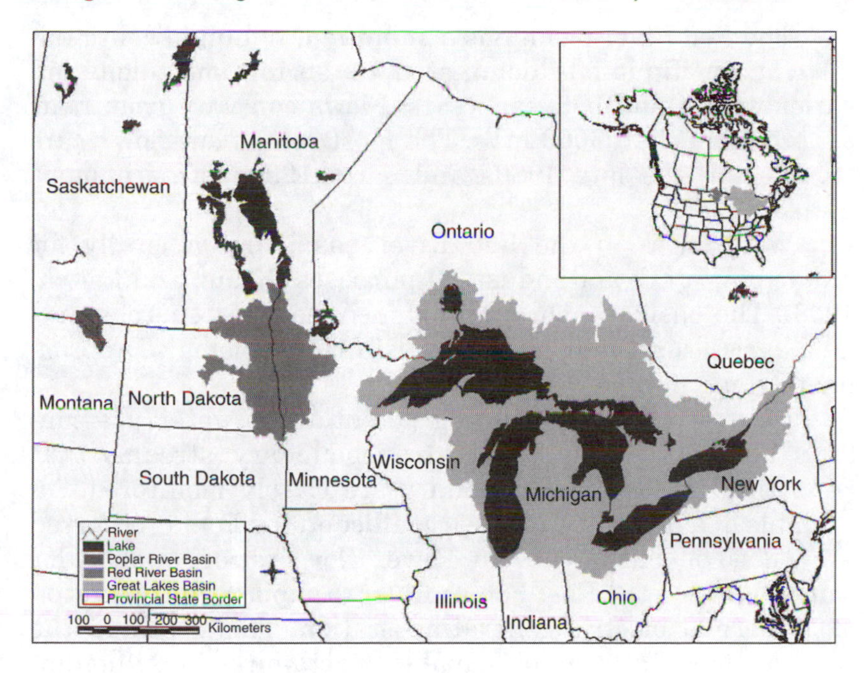

Figure 2 Poplar, Red River, and Great Lakes basins.

shifted to March instead of April (Environment Canada, 2002).

The sparsely populated Poplar basin is primarily agricultural. The main water uses are irrigated agricultural land in Montana and cooling of a thermal electric generating station in Saskatchewan.

The Red River (Figure 2), known officially as the Red River of the North in the United States, originates as the Otter Tail and Bois de Sioux Rivers in Minnesota and South Dakota, respectively, and flows northward into Canada, forming the boundary between Minnesota and North Dakota. In Canada, the river flows into Lake Winnipeg, the ninth largest lake in the world, and then to Hudson Bay via the Nelson River. Almost 90% of the 116,500 km^2 basin, exclusive of the Assiniboine River, lies in the United States, largely in North Dakota and Minnesota.

The Red River basin has a subhumid to humid continental climate. Runoff is dominated by spring snowmelt and varies within and between years. Flows can vary from near zero to more than 3000 m^3/s. The most recent low-flow years occurred in the late 1980s and early 1990s (Environment Canada, 2002).

Water uses in the Red River basin are generally for municipal, rural, and industrial purposes (Krenz and Leitch, 1998). The basin supports an extensive and prosperous dryland agricultural industry. The basin is also home to approximately one million people.

Because of the continuing potential for water use conflicts along the Canada–U.S. transboundary, streams that cross the international boundary are closely monitored. An example of a historic water use conflict on the Poplar occurred in 1972. The Saskatchewan Power Corporation applied for water rights on the East Poplar River to support the operation of a thermal electric generating station. This required the construction of a reservoir capable of retaining 40 million m^3 of water, roughly 35% of the mean annual flow in the basin.

As a result of an IJC investigation (IJC, 1978), the waters of the Poplar River are apportioned between Canada and the United States. The IJC's apportionment recommendations (adhered to but not formally accepted by the two countries) call for the waters of the basin to be divided equally between the two countries but for an asymmetric distribution among the three tributaries to accommodate the cooling water requirement on the East Poplar.

The transboundary effects of the generating station are monitored by a binational committee (Poplar River Bilateral Monitoring Committee [PRBMC]). These effects relate to groundwater and to surface water quality. Several water quality parameters are monitored, particularly boron and total dissolved solids, and compared to water quality objectives designed to prevent harm to existing water uses in the United States (PRBMC, 2002).

Recently, flooding has been the key issue on the Red River basin (IJC, 2000), but the basin has experienced droughts as well. Although no formal international agreements exclusively cover the Red River (Bruce et al., 2003), concerns that untreated or poorly treated municipal and industrial effluents entering the Red River from the United States may be impairing water uses in Canada led to a limited set of agreed-upon water quality objectives at the international boundary in 1969. The IJC established a board, now known as the International Red River Board (IRRB), to administer the objectives and report to governments (IRRB, 2002).

Canada has also expressed great concern regarding the potential effects, in particular on the aboriginal and commercial fishery of Lake Winnipeg, of proposed North Dakota water projects such as the Garrison Diversion Unit on Canadian waters. The proposed projects divert water from the Missouri basin to the Hudson Bay basin, and the concern is two-fold: degradation of water quality and the introduction of invasive species (IJC, 1977; Kellow and Williamson, 2001).

Multi-decadal droughts occurred in the region before European settlement. Lake salinity records covering 2000 years are used as proxy drought data for Moon Lake, North Dakota (Liard et al., 1996). The records indicate that multi-decadal droughts occurred before AD 1200 but not in subsequent years. Further evidence (Sauchyn and Beaudoin, 1998) indicates that the 20th century was relatively benign climatologically and confirms that decade-long droughts occurred before European settlement. Rannie (1999) identifies only two 3-year droughts in the pre-instrumental historical record (1793–1870) for the Red River: 1816–18 and 1862–64.

In more recent times, droughts have affected water use. For example, during a prolonged dry period beginning in 1987, groundwater pumping was used to mitigate decreased volumes and degraded water quality of the Cookson reservoir in the East Poplar River (PRBMC, 2002). The societal impacts of the recent droughts (1987–88 and 2000–02) have not been thoroughly documented for the Poplar and Red basins. The flows of both streams were severely reduced during the late 1980s, with effects on agricultural production and municipal and industrial water supplies, and decreased assimilative capacity.

Although both the Poplar and Red basins are susceptible to drought events, no comprehensive drought management plans exist for either basin. The primary public and institutional response to Red River water quantity problems is to augment supply through importation from the Missouri basin (*Fargo Forum*, 2003).

Future climate change will have an impact on available water in both the Poplar and Red basins. The Poplar basin is projected to have a 2–4°C temperature increase by the 2050s from the 1961–1990 averages. By the 2080s, the basin's temperature is projected to increase by 3–6°C. The CGCM2 and CSIRO model runs indicate that the majority of the warming will be in the winter and spring whereas the HADCM3 indicates that the summer and fall seasons will have the largest amount of warming. Precipitation amounts are expected to increase on an annual basis in both the 2050 and 2080 periods. However, the summer season is projected to have less precipitation than what was received in the 1961–1990 period (Canadian Institute of Climate Studies, 2003).

The Red River basin is projected to have a temperature increase of 2–4°C in the 2050s and 3–7°C in the 2080s. Annual precipitation values are expected to be below the 1961–1990 values for the 2050 and 2080 periods for the CGCM2 and CSIRO model runs and close to the 1961–1990 values for the HADCM3 model. The largest decrease in precipitation is projected to be in the summer (Canadian Institute of Climate Studies, 2003).

The result of these climatic changes will be that the prairies, on average, will likely experience significantly reduced spring runoff with the possibility of more severe summer rainfall events. Decreased water availability will lead to greater likelihood of transboundary conflicts concerning water use.

There are a number of potential consequences for the Poplar basin. First, on-farm water demands could increase. Canadian demands could match or exceed the entitlement under the current apportionment arrangement. Irrigation water demands in the United States, as well as other on-farm demands, would likely increase. Water use in the United States may be further affected by Montana water rights administration. Under the principle of prior apportionment (Lucas, 1990; Wolfe, 1996), the water rights associated with the Fort Peck Indian Reservation are senior rights, and the needs of the reservation (IJC, 1978) must be met before those of irrigators between the reservation and the international boundary. The response to such a situation could be to attempt to renegotiate the apportionment arrangement with Canada at a time when Canadian farmers are also facing water shortages.

Second, a suite of potential problems is associated with the generating station in Canada. Under drought conditions, the quantity of cooling water in Cookson reservoir would be insufficient to cool both units at the generating station. Therefore, as reservoir water temperatures increase, the plant must be de-rated. Further, during periods of low inflow, the water quality in Cookson reservoir steadily degrades through evaporation until it may exceed the water quality objective for total dissolved solids, and thus affect use of the water released to the United States.

The onset of prolonged droughts in the Red River basin as projected with climate change scenarios will lead to increased water demands for municipal and industrial purposes. The communities in the Red River basin in the United States are abundant water users, with per capita use in Fargo much greater than that in Winnipeg or desert cities such as Tuscon (*Fargo Forum*, 2001; D. Griffen, personal communication, 2001). One could assume that the effects of drought may be met through water conservation.

There may also be a problem with the lower flows in the Red River and its ability to assimilate municipal wastewater, especially considering the large population along the river. The quantity of water may become an issue even with urban conservation measures. At present, irrigation water demand in the basin is minor, but an increase in irrigation development on account of drought would have a profound effect on water supplies. Irrigation return flows may also affect water quality.

An increase in water demand of that magnitude would lead to Canadian pressure to formally apportion the waters of the Red River between the two countries, as is done for other prairie basins. This task would be complex of itself because North Dakota and Manitoba administer water rights on the basis of prior apportionment, but Minnesota uses riparian water law (Lucas, 1990). More important, increased water demands on the Red River may well lead to increased pressure to divert water from the Missouri River to meet agricultural and other needs in North Dakota, rather than curtailing water uses as would be required under apportionment or conservation measures. This pressure to divert water raises concerns over degradation of water quality and the introduction of invasive species.

C. Great Lakes Basin

The Great Lakes—Superior, Michigan, Huron, Erie, and Ontario—contain approximately 18% of the world's freshwater (Government of Canada and the U.S. Environmental Protection Agency, 1995). This leads to a perception of an extraordinary abundance of water. Yet, only 1% of the water in the Great Lakes is renewable on an annual basis; the rest is a legacy of deglaciation (Gabriel and Kreutzwiser, 1993).

The Great Lakes basin (including St. Lawrence River to Trois Rivières) contains the Canadian provinces of Quebec and Ontario and eight U.S. states (Minnesota, Wisconsin, Michigan, Illinois, Indiana, New York, Ohio, and Pennsylvania) (Figure 2). It is home to roughly one-quarter of the Canadian population and one-tenth of the U.S. population. The lakes play an important role in Canada and the United States, providing resources and opportunities important to many sectors of the region, including agriculture, forestry, fisheries, recreation and tourism, domestic and industrial water sources, navigation, and hydropower.

Before about 1950, a number of water diversions into, out of, and within the Great Lakes–St. Lawrence system were initiated and many continue today. The Long Lac and Ogoki diversion brings water into the basin from the Hudson Bay watershed and has a mean flow of 159m^3/sec. The diversion

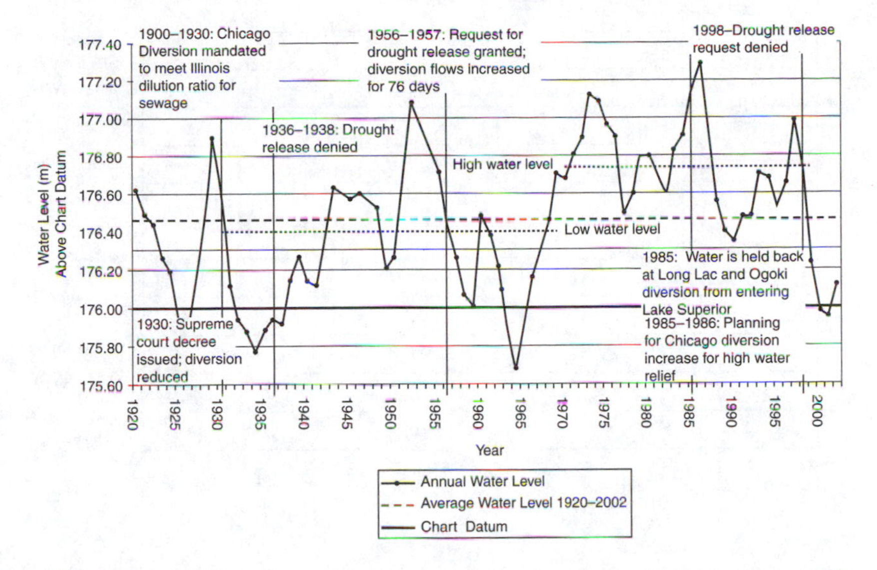

Figure 3 Timeline of Great Lakes diversions and Lake Michigan–Huron annual average water levels for 1920–2002. (Modified from Changnon and Glantz, 1996.)

at Chicago, transferring water from Lake Michigan to outside of the Great Lakes basin, is 91m^3/sec. The Welland Canal (9400 cfs) and New York State Barge Canal (1070 cfs) transfer water from Lake Erie to Lake Ontario (Cuthbert and Muir, 1991).

A period of low levels occurred in the Great Lakes between the 1930s and the 1960s, followed by a period of high water levels between the 1970s and the late 1990s (see Figure 3).

In 1998, drought conditions began within the Great Lakes basin. Although water levels in Lakes Michigan-Huron were quite high, the annual Palmer Drought Severity Index (PDSI) was beginning to dip below the threshold of –3 to indicate a severe drought was imminent (Figure 4). The following year, 1999, water levels began to drop, ending the period of high water levels that had been characteristic of the lakes since the early 1970s. In 2000 and 2001, water levels were significantly below average, reaching levels that were last seen during the drought of the 1960s. The PDSI for 2000

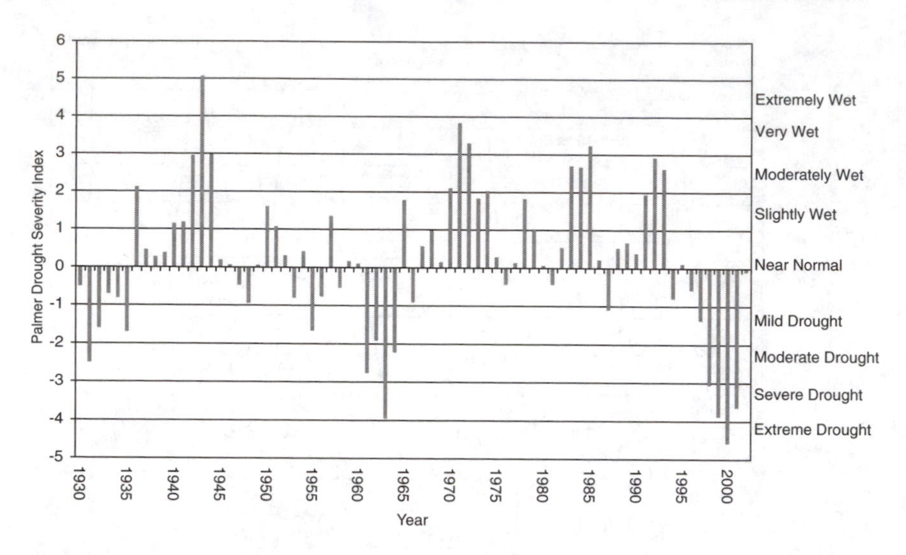

Figure 4 Annual Palmer Drought Severity Index for Gore Bay, Ontario, 1930–2002.

reached extreme drought levels and gradually receded to severe drought levels in 2001. Water levels reached borderline extreme negative levels in 2002, and the PDSI returned to a normal level.

Drought conditions can influence the viability and functionality of the Great Lakes basin and affect many sectors. Table 1 lists several examples.

Many Great Lakes provinces and states have developed state- and province-wide drought management and mitigation plans with varying degrees of detail: Ontario (2000), Illinois (1983), Indiana (2002), Michigan (1988), Minnesota (1993), New York (1982), Pennsylvania (2001), and Ohio (1994) (Commonwealth of Pennsylvania—Department of Environmental Protection, 2001; Illinois State Water Plan Task Force, 1983; Indiana Department of Natural Resources, 2000; Michigan Department of Natural Resources, 1988; Minnesota Department of Natural Resources, 1993; New York State Drought Management Task Force, 1982; Ohio Emergency Management Agency, 1994; Ontario Ministry of Natural Resources et al., 2001). (The drought plans for the U.S. states listed are better

Table 1 Drought Impacts in the Great Lakes

Sector	Impact
Hydro	• Decrease in hydroelectricity production. • Loss of revenue to power producers. • Increase in power demands from users.
Navigation/dredging	• Vessels carry reduced loads in order to clear ports and channels; must make extra trips to transport goods. Profits are decreased. • Vessels being kept out of service because they are not able to navigate properly. • Vessels prohibited from passing through certain canals. • Increase in dredging costs for marinas.
Recreation/wildlife	• Increased dangers for boaters; obstructions normally below the water surface are emerging (e.g., logs and rocks). • Recreational boaters have a difficult time entering and exiting marina boat slips. • Shoreline receding. • Beaches closed; bacteria aggravated by lower water levels.
Ecosystem	• Sports clubs, conservation authorities, and government agencies advising fishermen to avoid fishing, because the fish are subject to high rates of mortality due to stress. • Hundreds of fish have been killed in many areas; many stranded in shallow puddles within dried up streams and wetlands. • Fish have trouble spawning. • Ducks have perished because of a dry spring. • An increase in forest fires. The fire season began earlier than usual in some places.
Property interests	• Nonessential watering bans issued. • Rural residents in some areas have dry shoreline wells. • Lower levels lead to new islands emerging and the expansion of existing ones, creating issues concerning wildlife, property rights, and land management. • Reservoir levels very low.

Table 1 Drought Impacts in the Great Lakes (continued)

Sector	Impact
Agriculture	• Drought destroys many acres of crops. • Early harvest for many farmers. Many crops are lost because they were scorching in the fields. • The quality of many crops is less than desirable. • Many farm wells go dry. Water is either hauled in or fetched by the farmers themselves at municipal taps. Farmers sell livestock in many cases to deal with the lack of water, rather than go into future debt raising them.

Sources: Associated Press (2002), Brotton (1995), Anonymous (1999a), Anonymous (1999b), Anonymous (2003), Churchill (1998), Diebel (1999), Ferris (1999), Gabriel and Kreutzwiser (1993), Hughes (1999), Ladan (1998), Longbottom (1998), Marr (1998), Nolan (1999), Romahn (1998), Schuck (1998), *Toronto Star* (1998), van Rijn (1998), van Rijn (1999), Volmers (2001), Younglai (2001).

defined as drought response plans rather than drought management plans. They certainly have little emphasis on drought mitigation.) Many plans are 10 years old or older. Common components include assessment tools (drought triggers), water conservation and drought contingency planning legislation and policy, water allocation or supply methods, public education, and emergency response procedures.

Most jurisdictions focus on increasing resilience to drought by improving conservation, and securing supplies from groundwater, reservoirs, and small lakes. The Michigan plan explicitly does not support diversions outside the basin. It sees diversions as creating an unalterable dependency on Great Lakes water by out-of-basin users, which threatens the continued availability of water for Great Lakes uses. The New York plan identifies the southeastern portion of the state as vulnerable to critical water shortages under extreme drought. Although the primary focus is securing local water sources and conservation, two potential long-term, $22m^3$/sec water transfer projects were identified. They include the New York Barge Canal system with the Finger Lakes and Lake Champlain diverted into the Hudson River and a Great Lakes diversion to the Hudson River via the Black River system.

Diversion of water out of the Great Lakes is an extremely sensitive inter-jurisdictional issue. In the United States, legal precedent in the 1980s on water controversies in Wyoming, Idaho, Oregon, and Colorado has made interbasin transfer legally possible. The 1985 Great Lakes Charter is an institutional framework designed to deal with requests for diversion and export of water. Although not a legally binding document, its signatories (all seven Great Lakes states and two Canadian provinces) commit to consult on major new or increased diversion or consumptive use of Great Lakes basin water. The proposed diversion or export will be disallowed if significant adverse impacts to lake levels, in-basin uses, and the Great Lakes are expected.

Population growth, urbanization, security and quality of water supply, power generation, transportation, international markets for water, and water levels as well as drought are key drivers in creating water controversies. During the low-water period in 1965, downstream and upstream interests and various water users were in conflict. For example, Montreal Harbour interests charged that Ontario Hydro was holding back water for future use and affecting navigation on the St. Lawrence River. Meanwhile, boating and shoreline interests on Lake Ontario thought levels in Montreal Harbour were being maintained at their expense and Ontario Hydro was discharging more water downstream to produce power than what was entering the lake. Although the IJC regulation plan for managing the outflows from Lake Ontario was developed to accommodate sequences of water supplies from 1860 to 1954, the 1960s drought and the high supplies of the 1970s and 1980s challenged the ability of the plan to accommodate all requirements of multiple interests.

Many communities view the waters of the Great Lakes as a valuable resource. Lowell, Indiana, just outside the Great Lakes watershed limits, proposed a pipeline to access Lake Michigan water to improve drinking water quality and expand municipal services, but it was vetoed by Michigan in 1992 as contrary to the Great Lakes Charter. Many small local out-of-basin transfers of water have been proposed, but all have been denied because they set a precedent and have potential cumulative impact on the lakes.

In 1998, the Province of Ontario granted a permit to take water to the Nova Corporation for the removal of Lake Superior water (up to 600 million liters per year over a 5-year period) for export to Asian markets via large tanker ships. An outcry ensued, including the U.S. government, environmental organizations, and native groups (*National Post*, 1998). The permit was later rescinded, but the incident initiated a process to amend the Great Lakes Charter with a legally binding annex to prohibit withdrawals of large quantities of water without notice and consent; the annex also required the development of standards for review of proposals, public participation, and dispute resolution. Bulk water transfer is a serious issue, where project approval could trigger the interpretation that water is a commercial commodity subject to unrestricted trade under international agreements (Bruce et al., 2003).

In 1900, Chicago constructed a canal to divert Lake Michigan water to drain the sewage down the Illinois River and prevent water-borne disease outbreaks. During the 1930s, the Illinois River was developed as a navigation link between Chicago and the Mississippi River. The conflict between Illinois and other states over the continuously increasing diversion and its perceived impact on lake levels was taken to the U.S. Supreme Court, which set the diversion at 91m^3/sec. Drought has precipitated a number of out-of-basin transfer proposals. In 1936–1938, record low flow in the Mississippi River motivated Illinois and states along the river to request an increase in the diversion; the Supreme Court refused. A special 76-day increase in the flow to the Mississippi River was allowed during the drought of 1953–1956. During the 1988 drought, Illinois requested a diversion of 283m^3/sec for 100 days for navigation but was defeated by other Great Lakes states and Canada. During high water levels, the Corps of Engineers was authorized in 1976 to study the effects of a diversion increase to help alleviate high water levels (Changnon, 1989; Changnon and Glantz, 1996) (Figure 3).

Climate change may result in less water in the Great Lakes Basin, including significant reductions in groundwater levels, streamflow, and lake levels (Bruce et al., 2003; Kunkel and Changnon, 1998; Lofgren et al., 2002; Mortsch and Quinn, 1996; Mortsch et al., 2000).Also, within the North American context, some regions may have less reliable and reduced water supplies in the future. Climate change effects on water and transboundary issues could unfold in the following contexts:

1. Inland communities with diminishing supplies from groundwater and streams and declining water quality will want access to Great Lakes water.
2. Regulation plans were not designed to accommodate the low net basin supplies and connecting channel flows with climate change scenarios (Lee and Quinn, 1994). It will be challenging to balance the many interests if there is insufficient water.

3. The Niagara River Treaty allows equal apportionment of Niagara River flows between Canada and the United States for hydroelectric generation. As of 2000, the treaty can be reopened for negotiation. Should the amount of water in the Niagara River become seriously reduced because of climate change, apportionment could become a point of negotiation between the two countries (Bruce et al., 2003). Canada may want credit for the Ogoki-Long Lac diversion at more generation sites and the Chicago diversion debited to the U.S. side.
4. Although potential water demands from the U.S. southwest are more apparent and the Chicago diversion could be expanded, the demand from cities such as New York and Philadelphia may be an additional diversion pressure (Changnon, 1994; Changnon and Glantz, 1996).

Increasing numbers and fractiousness of inter-jurisdictional conflicts over the quality and availability of water resources seems highly probable (Bruce et al., 2003). It is uncertain whether the Great Lakes Charter will be a successful instrument for the Great Lakes states and provinces to protect their shared water resources from diversion pressures.

IV. CONCLUSIONS

The regional case studies have shown that past water shortages caused by drought conditions can affect the quantity and quality of water available to multiple users. The preparedness of these watersheds for past and future drought conditions is mixed. Although both the Poplar and Red River basins are susceptible to droughts, no comprehensive drought management plans exist for either watershed. By comparison, many provinces and states in the Great Lakes basin have developed state- and province-wide drought management plans. Although most jurisdictions focus on increasing resilience to drought by improving water conservation and securing additional water supplies, there is no indication that these drought management plans have had a significant effect on reducing the region's vulnerability to drought.

Demands for water diversions or intra-basin transfers of water are expected to grow in the future. In developing future water management policies, water managers can no longer assume that the climate will remain constant. Climate change scenarios suggest longer and warmer growing seasons, reduced water supplies, and increased water demands. The frequency and severity of climate extremes such as droughts are also expected to change in the future. Sustainable management of water resources in these watersheds will require water managers to choose areas of priority within the constraints of existing supplies, legislation, and international agreements.

Although there is a growing consensus on the potential implications of climate change for water resources throughout the world (Arnell et al., 2001), including North America (Lettenmaier et al., 1999; Natural Resources Canada, 2002), linking such concerns to water management requires additional mechanisms to explore and test, in a virtual sense, specific options that would work within the context of regional development plans.

REFERENCES

Arnell, N; Liu, C; Compagnucci, R; da Cunha, L; Howe, C; Hanaki, K; Mailu, G; Shiklomanov, I; Stakhiv, E. Hydrology and water resources. In: JO McCarthy, O Canziani, N Leary, D Dokken, K White, eds. *Climate Change 2001: Impacts, Adaptation, and Vulnerability. Contribution of Working Group II to the Third Assessment Report of the Intergovernmental Panel on Climate Change*. London: Cambridge University Press, 2001.

Associated Press. Great Lakes levels should be higher this summer. *The Record*, March 9, A11, 2002.

Bonsal, B; Chipanshi, A; Grant, C; Koshida, G; Kulshreshtha, S; Wheaton, E; Wittrock, V. Saskatchewan Research Council reports. In: E Wheaton, S Kulshreshtha, V Wittrock, eds. Canadian Droughts of 2001 and 2002: Climatology, Impacts and Adaptations. Prepared for Agriculture and Agri-Food Canada. Saskatchewan Research Council, Saskatoon, Saskatchewan, 2003.

Bonsal, B; Koshida, G; O'Brien, EG; Wheaton, E. Droughts. In: Threats to water availability in Canada (pp. 19–25). NWRI Scientific Assessment Report Series No. 3 and ASCD Science Assessment Series No. 1. Environment Canada, National Water Research Institute, Burlington, Ontario, 2004.

Brotton, J. Causes and impacts of 1960s low water levels on Canadian Great Lakes interests. Environmental Adaptation Research Group, Atmospheric Environment Service, Burlington, Ontario, 1995.

Bruce, JP; Martin, H; Colucci, P; McBean, G; McDougall, J; Shrubsole, D; Whalley, J; Halliday, R; Alden, M; Mortsch, L; Mills, B. *Climate Change Impacts on Boundary and Transboundary Water Management*. Ottawa: Natural Resources Canada, 2003.

Canadian Institute of Climate Studies. Global Climate Model Scenarios [Web Page]. Accessed October 15, 2003. Available at *http://www.cics.uvic.ca/scenarios/index.cgi?Scenarios*, 2003.

Anon. High winds stop ships. *Guelph Mercury*, November 16, A6, 1999a.

Anon. Lake Ontario water levels expected to increase. November 22, Quebec–Ontario regional general news, 1999b.

Anon. Profits unplugged at OPG. *Toronto Star*, August 7, D10, 2003.

Changnon, SA. The drought, barges, and diversion. *Bulletin of the American Meteorological Society* 70:1092–1104, 1989.

Changnon, SA; ed. The Lake Michigan diversion at Chicago and urban drought: Past, present, and future regional impacts and responses to global climate change. Report to the Great Lakes Environmental Research Laboratory, CCR-36, NOAA, Washington, D.C., 1994.

Changnon, SA; Glantz, MH. The Great Lakes diversion at Chicago and its implications for climate change. *Climatic Change* 32:199–214, 1996.

Chiotti, Q; Morton, I; Ogilvie, K; Maarouf, A; Kelleher, M. Towards an adaptation action plan: Climate change and health in Toronto–Niagara region. A pollution probe report, Toronto, 2002.

Churchill, J. Farmers pin hopes on forecast: Extra dry July leaves reservoirs at their lowest level in decades. *The Record*, August 6, A1, 1998.

Cohen, S; Kulkarni, T; eds. Water Management & Climate Change in the Okanagan Basin. Environment Canada & University of British Columbia. Project A206, submitted to the Adaptation Liaison Office, Climate Change Action Fund, Natural Resources Canada, Ottawa, 2001.

Cohen, SJ; Miller, KA; Hamlet, AF; Avis, W. Climate change and resource management in the Columbia River Basin. *Water International* 25(2):253–272, 2000.

Cohen, S; Miller, K; Duncan, K; Gregorich, E; Groffman, P; Kovacs, P; Magana, V; McKnight, D; Mills, E; Schimel, D. North America. In: J. McCarthy, O Canziani, N Leary, D Dokken, K White, eds. *Climate Change 2001, Impacts, Adaptation and Vulnerability. Contribution of Working Group II to the Third Assessment Report of the Intergovernmental Panel on Climate Change* (pp. 735–800). London: Cambridge University Press, 2001.

Cohen, S; Neale, T; eds. Expanding the dialogue on climate change and water management in the Okanagan Basin, British Columbia. Environment Canada, Agriculture and Agri-Food Canada and University of British Columbia. Project A463/433. Submitted to the Adaptation Liaison Office, Climate Change Action Fund, Natural Resources Canada, Ottawa, 2003.

Cohen, S; Neilsen, D; and Welbourn R; eds. Expanding the dialogue on climate change and water management in the Okanagan basin, British Columbia. Final Report, January 1, 2002 to June 30, 2004. Environment Canada, Agriculture & Agri-Food Canada, and University of British Columbia. Project A463/433. Submitted to Climate Change Impacts and Adaptation Program, Natural Resources Canada, Otawa, 2004.

Commonwealth of Pennsylvania—Department of Environmental Protection. *Fact Sheet—Drought Management in Pennsylvania.* Harrisburg, PA: Department of Environmental Protection, 2001.

Cuthbert, D; Muir, T. The Great Lakes basin and existing diversions. In: M Sanderson, ed. *Water Pipelines and Diversions in the Great Lakes Basin* (pp. 1–8). Waterloo, Ontario: Department of Geography Publication Series, 1991.

Diebel, L. Low water levels a mixed blessing. *The Hamilton Spectator*, November 9, A5, 1999.

Embley, E; Hall, K; Cohen, S. Background. In: S Cohen, T Kulkarni, eds. Water management & climate change in the Okanagan Basin. Environment Canada & University of British Columbia. Project A206. Submitted to the Adaptation Liaison Office, Climate Change Action Fund, Natural Resources Canada, Ottawa, 2001.

Environment Canada. *HyDat CD-ROM Version 2000*. Downsview, Ontario: Environment Canada, 2002.

Epstein, PR; Defilippo, C. West Nile virus and drought. *Global Change and Human Health* 2(2):105–107, 2001.

Etkin, D. Climate change and extreme events. In: N Mayer, W Avis, eds. *National Cross-Cutting Issues* (pp. 31–80), Volume VIII of the Canada Country Study: Climate Impacts and Adaptation. Toronto: University of Toronto Press, 1997.

Fargo Forum. Will Fargo water supply last? Editorial, July 30, 2001.

Fargo Forum. Eventually a drought will come. Editorial, October 7, 2003.

Ferris, A. Sunny weather leaves Puslinch lake low, residents high and dry. *The Record*, July 28, B5, 1999.

Gabriel, A; Kreutzwiser, R. Drought hazard in Ontario: A review of impacts, 1960–1989, and management implications. *Canadian Water Resources Journal* 18:117–132, 1993.

Government of Canada and the U.S. Environmental Protection Agency. *The Great Lakes: An Environmental Atlas and Resource Book*. Toronto: Government of Canada; and Chicago, Illinois: Great Lakes National Program Office, U.S. Environmental Protection Agency, 1995.

Hamlet, AF; Lettenmaier, DP. Effects of climate change on hydrology and water resources in the Columbia River Basin. *Journal of the American Water Resources Association* 35(6):1597–1624, 1999.

Hughes. R. Ships forced to carry lighter loads: Lower levels in the Great Lakes mean tight squeeze for vessels. *The Hamilton Spectator,* August 28, A13, 1999.

Illinois State Water Plan Task Force. Drought contingency planning. Special Report No. 3. Illinois Division of Water Resources, Department of Transportation, Springfield, IL, 1983.

Indiana Department of Natural Resources. *Indiana's Water Shortage Plan.* Indianapolis, IN: Indiana Department of Natural Resources—Division of Water, 2000.

International Joint Commission (IJC). Transboundary Implications of the Garrison Diversion Unit. An IJC Report to the Governments of Canada and the United States. Ottawa and Washington, D.C., 1977.

International Joint Commission (IJC). Water Apportionment in the Poplar River Basin. An IJC Report to Canada and the United States. Ottawa and Washington, D.C., 1978.

International Joint Commission (IJC). The International Joint Commission and the Boundary Waters Treaty. Ottawa and Washington, D.C., 1990.

International Joint Commission (IJC). Living with the Red. A Report to the Governments of Canada and the United States on Reducing Flood Impacts in the Red River Basin. Ottawa and Washington, D.C., 2000.

International Red River Board (IRRB). Third Annual Progress Report. Ottawa and Washington, D.C., 2002.

Kellow, RL; Williamson, DA. Transboundary Considerations in Evaluating Interbasin Water Transfers. Proceedings of the 2001 Water Management Conference, U.S. Committee on Irrigation and Drainage, 2001.

Koshida, G; Burton, I; Cohen, SJ; Cuthbert, D; Mayer, N; Mills, B; Mortsch, L; Slivitsky, M; Smith, J. Climate change: Practising adaptive management for sustainability of Canadian water resources. In: B Mitchell, D Shrubsole, eds. *Practising Sustainable Water Management: Canadian and International Experiences* (pp. 75–98). Cambridge, Ontario: Canadian Water Resources Association, 1997.

Koshida, G; Mills, B; Sanderson, M. Adaptation lessons learned (and forgotten) from the 1988 and 1998 Southern Ontario droughts. In: I Burton, M Kerry, S Kalhok, M Vandierendonck, eds. *Report of the Adaptation Learning Experiment* (pp. 23–35). Toronto: Environment Canada and Emergency Preparedness Canada, 1999.

Krenz, G; Leitch, J. *A River Runs North,* 2nd ed. Moorhead, MN: Red River Water Resources Council, 1998.

Kunkel, K; Changnon, S. Transposed climates for the study of water supply variability on the Laurentian Great Lakes. *Climatic Change* 38:387–404, 1998.

Ladan, M. Low Muskoka water levels heighten risk for boaters. *Toronto Star*, August 5, 1998.

Lee, DH; Quinn, FH. Modification of Great Lakes regulation plans for simulation of maximum Lake Ontario outflow. *Journal of Great Lakes Research* 20:569–582, 1994.

Lettenmaier, DP; Wood, AW; Palmer, RN; Wood, EF; Stakhiv, EZ. Water resources implications of global warming: A U.S. regional perspective. *Climatic Change* 43(3):537–579, 1999.

Liard, KR; Fritz, SC; Maasch, KA; Cumming, BF. Greater drought intensity and frequency before AD1200 in the Northern Great Plains, USA. *Nature* 384(12):552–554, 1996.

Lofgren, BM; Quinn, FH; Clites, AH; Assel, RA; Eberbardt, AJ; Luukkonen, CL. Evaluation of the potential impacts on Great Lakes water resources based on climate scenarios of two GCMs. *Journal of Great Lakes Research* 28:537–554, 2002.

Longbottom, R. Fish left without a place to swim. *The Hamilton Spectator*, November 7, N1, 1998.

Lucas, AR. *Security of Title in Canadian Water Rights*. Calgary: The Canadian Institute of Resources Law, Faculty of Law, University of Calgary, 1990.

Marr, L. Farms desperate for water: This has been one of the driest years on record in Canada. *The Hamilton Spectator*, December 17, A1, 1998.

Merritt, W; Alila, Y; Hydrology. In Cohen et al., 63–88, 2004.

Michigan Department of Natural Resources. *Drought Response Plan*. Lansing, MI: Office of Water Resources, Michigan Department of Natural Resources, 1988.

Miles, EL; Hamlet, AF; Snover, AK; Callahan, B; Fluharty, D. Pacific Northwest regional assessment: The impacts of climate variability and climate change on the water resources of the Columbia River Basin. *Journal of the American Water Resources Association* 36(2):399–420, 2000.

Minnesota Department of Natural Resources. *Drought Response Plan*. St. Paul, MN: Minnesota Department of Natural Resources—Division of Waters, 1993.

Moorhouse, J. Water crisis. *Penticton Herald*, July 31, 2003.

Mortsch, L; Quinn, F. Climate change scenarios for Great Lakes Basin ecosystem studies. *Limnology and Oceanography* 41(5):903–911, 1996.

Mortsch, L; Hengeveld, H; Lister, M; Lofgren, B; Quinn, F; Silvitzky, M; Wenger, L. Climate change impacts on the hydrology of the Great Lakes–St. Lawrence system. *Canadian Water Resources Journal* 25:153–179, 2000.

Mote, P; Canning, D; Fluharty, D; Francis, R; Franklin, J; Hamlet, A; Hershman, M; Holmberg, M; Gray-Ideker, K; Keeton, WS; Lettenmaier, D; Leung, R; Mantua, N; Miles, E; Noble, B; Parandvash, H; Peterson, DW; Snover, A; Willard, S. *Impacts of Climate Variability and Change, Pacific Northwest*. Washington, D.C.: National Atmospheric and Oceanic Administration, Office of Global Programs; and Seattle, WA: JISAO/SMA Climate Impacts Group, 1999.

National Post. Appeal over Lake Superior water pollution. November 28, A6, 1998.

Natural Resources Canada (NRCan). Water resources. In: *Climate Change Impacts and Adaptation: A Canadian Perspective. Climate Change Impacts and Adaptation Directorate*. Ottawa: Natural Resources Canada, 2002.

Neilsen, D; Smith, CAS; Koch, W; Frank, G; Hall, J; Parchomchuk, P. Impact of climate change on irrigated agriculture in the Okanagan Valley, British Columbia. Final Report, Climate Change Action Fund Project A087. Natural Resources Canada, Ottawa, 2001.

New York State Drought Management Task Force. *New York State Drought Preparedness Plan*. Albany, NY: New York State Drought Management Task Force—Department of Environmental Conservation, 1982.

Nolan, D. Low Great Lakes water levels continue; consecutive record-breaking warm years costing shippers and spoiling boaters' fun. *The Hamilton Spectator*, December 31, A2, 1999.

Ohio Emergency Management Agency. *Ohio Drought Response Plan*. Columbus, OH: Ohio Emergency Management Agency, 1994.

Ontario Ministry of Natural Resources, Ontario Ministry of the Environment, Ontario Ministry of Agriculture Food and Rural Affairs, Ontario Ministry of Municipal Affairs and Housing, Ontario Ministry of Economic Development and Trade, Association of Municipalities of Ontario, Conservation Ontario. Ontario Low Water Response, 2001.

Poplar River Bilateral Monitoring Committee (PRBMC). 2001 Annual Report to the Governments of Canada, United States, Saskatchewan and Montana, PRBMC, 2002.

Rannie, WF. Hydroclimate, Flooding and Runoff in the Red River Basin Prior to 1870. Geological Survey of Canada Open File 3705. Geological Survey of Canada, Ottawa, 1999.

Romahn, J. Nobody alive has seen such a year. *The Record*, September 21, D1, 1998.

Sauchyn, DJ; Beaudoin, AB. Recent environmental change in the southwestern Canadian plains. *The Canadian Geographer* 42(4):337–353, 1998.

Schuck, P. Park ducks die during dry spring. *The Record*, May 23, B1, 1998.

Shepherd, P. Adaptation experiences. In Cohen et al., 137–160 and Appendix, 213–216, 2004.

Toronto Star. High temperatures bad for muskies. *Toronto Star*, August 19, 1998.

van Rijn, N. Forest fire season a month early in north. *Toronto Star*, May 3, 1998.

van Rijn, N. Sizzling summer cooks record books. *Toronto Star*, July 31, 1999.

Volmers, E. Residents hit water to beat heat. *The Cambridge Reporter*, August 9, A3, 2001.

Watershed News. Low water in regional streams. Okanagan Nation Alliance—Fisheries Department, Westbank, B.C. September 12, 2003a.

Watershed News. Low water in regional streams. Okanagan Nation Alliance—Fisheries Department, Westbank, B.C. October 20, 2003b.

Wolfe, ME. *A Landowners Guide to Western Water Rights.* Bozeman, MT: The Watercourse, Montana State University, 1996.

Younglai, R. Residents asked to cut water use. *The Cambridge Reporter*, August 10, A1, 2001.

12

Drought and Water Management: Can China Meet Future Demand?

ZHANG HAI LUN, KE LI DAN,
AND ZHANG SHI FA

CONTENTS

I. PHYSIOGRAPHIC CONDITIONS AND WATER RESOURCES

A. Physiographic and Climate Conditions

1. Physiography

China is located in the southeastern part of the Asian continent. Its total area is 9.6 million km^2, which accounts for $^1/_{15}$ of the total land area of the earth. Topographically, China is divided from east to west into three areas consisting of plains, plateaus, and high mountains, which form a slope inclining toward the Pacific Ocean, with all major watercourses flowing from west to east. The highest area is the Qinghai-Tibet plateau, with an elevation higher than 4000 m and numerous mountains, valleys, and lakes. It is the source area of the Yangtze and Yellow rivers. The area to the north of the Qinghai-Tibet plateau and the eastern part of Sichuan Province, with an elevation of 1000–2000 m, constitutes the second area of China. It is composed entirely of plateaus and mountains. The Daxing'an, Taihang, and Wushan mountains and the area to the east bordering the Yunnan-Guizhou plateau up to the coastal areas of China constitute the third area, where the hills and plains crisscross each other. Most hills are less than 1000 m, and the elevation of the coastal plain areas is less than 50 m.

The low coastal plains and hills of eastern and southern China make up 41% of the total area of the country and are densely populated, with about three-quarters of China's population. The principal plains are the northeast plain, the north China plain, and the middle and lower Yangtze River basin plain.

2. Climate and Precipitation

The vast eastern area and most of the south are affected by the eastern Asia monsoon climate. During the summer season, these areas are affected by the oceanic air current, and in winter the continental air current prevails. This results in dry winters and wet summers. The summer monsoon plays an important role in the formation of rainfall for various regions of the country. About 60–80% of the total precipitation falls during the 4 months of the rainy season.

Rainfall is unevenly distributed throughout the country, ranging from an annual average rainfall of 2000 mm at the southeast coast to 200–400 mm in the northwest. This uneven spatial and temporal distribution has caused repeated natural calamities in China. The average annual precipitation in the mainland has been estimated at 648 mm; that is equivalent to a volume of 6189 billion m^3 (Department of Hydrology, Ministry of Water Resources, 1992).

B. Water Resources

1. Surface Water

The average annual runoff of all the rivers in China has been estimated at 2,711 billion m^3, equivalent to the mean runoff depth of 284 mm. China's total annual runoff ranks sixth in the world, after Brazil, the former U.S.S.R., Canada, the United States, and Indonesia. However, the per capita amount of runoff—estimated at 2,134 m^3/year—is only about a quarter of the world mean. The availability of water per hectare of cultivated land is estimated at 28,000 m^3, which is about two-thirds of the world average.

Water resources in China are characterized by uneven regional distribution (Table 1). The runoff of the Yangtze River and the river systems in the south accounts for 81% of the total runoff in the country, whereas the runoff of the large rivers to the north of the Yangtze River is only 14.4% of the total runoff.

2. Groundwater

The groundwater in China is classified as two types, based on the recharge characteristics. Shallow aquifers hydrologi-

Table 1 Surface Water Resources of China (by Region)

No.	Region/Basin	Drainage Area (km²)	Mean Surface Annual Runoff	
			mm	10^9m³
I	Northeastern	1,248,445	132.4	165.3
II	Hai He-Luan He basin	318,161	90.5	28.8
III	Huai He basin	329,211	225.1	74.1
IV	Yellow River basin	794,712	83.2	66.1
V	Yangtze River basin	1,808,500	526.0	951.3
VI	Southern	580,641	806.9	468.5
VII	Southeastern	239,803	1,066.3	255.7
VIII	Southwestern	851,406	687.5	585.3
IX	Interior basins	3,374,443	34.5	116.4
	National total	9,545,322	284.1	2,711.5

Source: Department of Hydrology, Ministry of Water Resources (1992).

cally and hydraulically interconnect with river flows and recharge by local precipitation. These aquifers serve as the major source of base flow of the rivers. In the low-lying areas of China, groundwater from shallow aquifers is widely used for irrigation and domestic water supply. It is renewable and is replenished yearly during the rainy season. The deep aquifers are formed over a long period of time and depend very little on precipitation and surface runoff.

In mountainous areas, groundwater from subsurface deposits and upstream flows are the major recharge sources. The total volume of groundwater in the mountainous areas throughout the country has been estimated at 676.2 billion m³.

3. Water Quality

The silt concentration is critically high in many rivers of China. Approximately 3.5 billion tons of sediment annually is transported to rivers from the mountains and hilly areas, of which about 60% discharges into the sea, while the remaining 40% deposits in the river courses in the middle and lower reaches, as well as in lakes and reservoirs. During flood periods, the sediments are transported to the lower reaches of the rivers, where they are deposited in the flood plains or diverted into the irrigation areas. With its annual transportation of about 1.6 billion tons, the Yellow River ranks highest among the world's largest rivers in terms of sediment transport. It accounts for 50% of the total river sedimentation of China; the annual runoff accounts for only 5% of the total.

Since the end of the 1970s, water quality has deteriorated because of arbitrary disposal of wastes in rivers. Industrial wastes and city sewage are the main point sources, whereas most of the nonpoint sources, mainly comprising pesticides and chemical fertilizer, originate from the agriculture sector. The comprehensive assessment of water quality of rivers in China conducted in 2000 indicates that the water quality is acceptable for most users along only about 58.7% of the total length of rivers.

II. DROUGHT AND WATER SHORTAGE ISSUES

A. Perception of Drought and Water Shortage

Drought is generally viewed as a sustained and regionally extensive occurrence of below-average natural water availability, in the form of precipitation, runoff, or groundwater. Drought should not be confused with aridity, which applies to those persistently dry regions where, even in normal circumstances, water is in short supply. Normally the consequences of droughts are felt most keenly in areas that are in any case arid (UNESCO-WMO, 1985). Furthermore, the adverse effects of drought are noticeable only in those areas that are inhabited, and the trend of drought impacts intensifies as human activities increase. If arid regions with sparse population are hit by drought, little or no impact will occur to humans or economies. But if the region is densely populated and highly developed, countermeasures have to be taken to cope with drought disasters. It is important to note that although drought may have many adverse effects on human activity, it is also true that many human activities aggravate drought (Zhang Hai Lun, 1997). Below are examples of drought in different regions of China with different natural and socioeconomic conditions.

In southern China, which has a humid climate, population pressure has led to an increase in the number of rice crops (from one to two) in many areas, which has resulted in frequent drought in some places. For example, in Anhui Province (located in the Yantze River and Huihe River basins), various measures have been taken, including changing from dry farming to growing rice, changing from one crop of rice to two, and applying more water to wheat. These activities led to a total water use in the agricultural sector of 14 billion m^3 in 1990, which is an increase of about 10 billion m^3 compared to the early 1950s.

The north China plain, located in a subhumid region with annual precipitation ranging from 600 to 800 mm, is historically a densely populated region with highly developed agriculture. Drought impacts have been increasingly critical since the 1970s, with small and medium rivers dry year round. Even the Yellow River becomes dry intermittently. The overdraft of groundwater has resulted in environmental degradation, and the frequency of drought events continues to increase.

In the Xinjiang and Ningxia Autonomous Regions, situated in the northwest arid zone, the dry climate does not lead to drought disasters because of stable water supplies from the snowfed rivers (Xinjiang) and the Yellow River (Ningxia).

B. Impact of Drought on Economic and Social Development

1. Major Natural Disaster in the Agricultural Sector

Drought is historically one of the major natural disasters affecting China, primarily the agricultural sector. Data for 1949–2000 show that drought affected an average of 21 million ha each year and disastrous droughts affected 8.9 million ha, which accounts for 21% and 8.9%, respectively, of total cultivated land in China (Zhang Shi Fa, 2002). During eight extraordinary dry years (1960, 1961, 1978, 1986, 1988, 1989, 1997, and 2000) in the period 1949–2000 the rate of drought-affected land to total cultivated land exceeded 20%. Average reduction of total grain output reached 5%, and was as high as 13% in 1961. The drought in 2000 hit many grain-producing regions; the total grain loss was estimated at 60 million tons, the highest loss in the past 51 years. The population affected by drought in China increased from 22 million in the 1950s to 90 million in the 1980s. In those extraordinary dry years (e.g., 1988), 132 million rural people were affected.

2. Increasing Water Shortage in Urban Areas

Water shortage in urban areas constitutes three types. The first one is due to insufficient water resources, meaning that urban, industrial, and environmental water use demands cannot be met by available water resources. The second type refers to those areas that lack adequate water supply facilities despite the fact that water is available for meeting those needs. The third type is a newly emerging one: the water resources of cities are heavily polluted. Many cities in humid coastal and delta areas are short of water because of water quality problems. In 1980, water shortage events extended to coastal areas, covering 21 cities in 11 provinces. In 1990, some 300 cities were short of water; this figure doubled in 2000. Unlike drought in rural areas, water needs of cities are met before the needs of other sectors, so that water shortage in cities is to some extent independent of an ordinary drought event. Only in severe drought events is the water supply in urban areas affected. It is difficult to compare water shortage events in cities in terms of their intensity, duration, and coverage.

3. Major Constraints on Development of Pasture Areas

Pasture areas covering 416 million ha are distributed in 12 provinces in north, west, and northwest China, with annual precipitation ranging from 50 to 400 mm. These areas are frequently stricken by droughts of long duration. In 2000, drought affected 78 million ha of grassland in pasture areas, causing severe losses in the pasture industry in Inner Mongolia and other provinces. In the period 1953–1991, there were 10 years during which more than 500,000 animals died of drought-related causes.

4. Drinking Water Supply Problems

In many rural areas, the distance between villages and their water sources can exceed 1–2 km. Difficult access to drinking water supplies has long been a critical issue in Chinese history, especially in loess plateaus, karst areas, and areas with poor vegetative cover and serious soil erosion. Because of population growth and the development of agricultural industry, water quality deterioration has accelerated in rural areas. As of 1995, drinking water supply was a problem for 73.3 million people in rural areas, accounting for 8% of China's rural population.

C. Impact of Drought on Environment

In arid and semiarid regions (annual precipitation less than 400 mm), local rainfall does not meet agricultural needs. In such cases, too many withdrawals from rivers upstream could directly affect the water environment and ecosystem downstream. Environmental issues due to drought or water shortage (mainly in north China) are discussed below.

1. Drying of Land, Lakes, and Rivers

In the north China plain, the annual runoff coefficient dropped from 0.2 in the 1950s to 0.1 in the 1980s because of too many withdrawals from rivers. Most of the small and medium rivers in the north China plain have become dry intermittently. Since 1972, even the Yellow River has frequently dried up in the spring and summer. Bai Yang Dian, the famous lake of the north China plain, dried up several times in the 1980s, leading to water pollution, sedimentation, and other adverse consequences for the environment and ecosystems. A similar situation occurred in the Tarim River in Xinjiang Autonomous Region and Shiyang River of Gansu Province in northwest China.

2. Environmental Degradation Due to Lowering of the Groundwater Table

Overdraft of groundwater in the north China plain and in urban areas in the Yangtze Delta led to a significant decline of the groundwater table and an expansion of cone coverage in these areas. By the end of the 1980s, the average decline of the groundwater table of shallow aquifers in the north China plain was 8–10 m, and the decline in cone areas, covering 27,000 km^2, reached as high as 20–30 m. The decline of the groundwater table has had critical environmental consequences. Land subsidence was found in cities like Beijing, Tianjin, and Taiyuan in the north and Shanghai, Suzhou, and Wuxi in the Yangtze River Delta.

Sea water intrusion in coastal areas is another environmental issue caused by the declining groundwater table. The affected area along the Bo Hai Bay for Liaoning, Hebei, and Shandong provinces totaled 1432 km^2.

3. Water Pollution

With rapid economic development in China, surface water quality has deteriorated in urban and coastal areas. In the late 1980s, polluted water directly discharged into rivers in the north China plain totaled 4.3 billion tons, but the total annual runoff is only 33.8 billion m^3. The ratios of polluted water to total runoff for Beijing, Tianjin, and Tangshan were even higher. In inland basins like Ta Li Mu River (Xin Jiang Autonomous Region), because of the decline of natural runoff and increase of irrigation return flow, the water is significantly alkalized, and the same situation occurred for many inland lakes in the northwest.

III. WATER USE TREND IN THE PAST HALF CENTURY

A. The Changing Rate of Water Use in Different Sectors and Administrative Regions

During the period 1949–2000, total annual water use in China ranged from 103.1 billion m^3 to 549.8 billion m^3 (Table 2), and the rate of increase was declining. It was high (5.2%) in 1949–1980, about two times the rate of increase in population growth (2.7%) for the same time period (Ke Li Dan, 2002). Since the early 1980s, China has maintained rapid economic development and has taken various important measures in water resources management, including initiating a planned water use program and a national water-saving program, reforming the water charge system, and instituting demand-oriented management for cities in north China. More importantly, China has successfully implemented the family planning program. As a result, the trend of a high increase in water use was restrained, and the average increase rate of water use during 1981–1993 dropped to 1.3%, lower than the population growth rate (1.4%) in the same time period. Since 1993, the rate of increase in water use has dropped further for a number of reasons, mainly the restructuring of the national economy, system reform, and the comprehensive water-saving activities. All these led to further declines in the increase rate: 0.8% in 1993–2000, which is much lower than the rate of population growth (1.1%).

In the same time period, the rate of increase in water use in 15 provinces, autonomous regions, and municipalities directly under the administration of central government reached zero growth (Table 3). This covers the developed coastal areas in the south and north and the water-scarce provinces in the north. The total water use of the country fluctuates around the mean of 550 billion m^3 in the seven consecutive years from 1997 to 2003.

TABLE 2 Annual Water Use in China (1949–2000)

Item	Year							Increase (%) 1993–2000	Average increase rate(%)
	1949	1959	1965	1980	1993	1997	2000		
Population (10^6)	550	650	740	980	1170	1250	1290	+9.4	+1.3
Annual water use (10^9 m^3)	103.1	204.8	274.4	443.7	519.8	556.6	549.8	+5.7	+0.8
Per capita water use (m^3)	187	315	370	452	444	445	430		
Urban water supply	0.6	1.4	1.8	6.8	23.7	25.6	28.4	+19.8	+2.8
(%)	0.6	0.7	0.7	5	4.5	4.4	5.2		
Industry	2.4	9.6	18.1	45.7	90.6	112.1	113.9	+25.7	+3.7
(%)	2.3	4.7	6.6	10.3	17.4	20.1	20.7		
Agriculture	100.1	193.8	254.5	391.2	405.5	419	407.5	+0.05	<0.1
(%)	97.1	94.6	92.7	88.2	78.0	75.5	74.1		
Irrigation (10^9 m^3)				357.1	344.0	360.6	346.7	0	0
Forestry, fishery (10^9 m^3)				13.1	37.7	31.4	31.7	–15.9	–2.3
Rural water supply (10^9 m^3)				21.0	23.8	27.9	29.1	+22.3	+3.2

Source: Ke Li Dan (2002).

TABLE 3 Water Use Indices for Administrative Regions in China

Item / Administrative Region	Total Water Use (10^9 m^3)		Increase of Water Use			Per Capita Water Use (2000) (m^3)	Water Use Efficiency (yuan GDP/m^3)				Water Use Structure (2000)		
	1993	2000	Increment	%	Rate of Increase		1997	2000	Increment	Rate of increase	Water supply	Industry	Agriculture
National	519.8	549.8	300	5.8	+0.8	430	13.7	16.4	2.7	6.6	1	2	6.6
Beijing	4.0	4.0	0	0	0	290	45	62.5	17.5	12.9	1	0.8	1.2
Tianjing	2.2	2.2	0	0	0	230	51.3	71.4	20.1	13.0	1	1	2.3
Hebei	20.9	21.6	0.7	3.3	+0.5	310	17.9	23.8	5.9	11.0	1	1.2	7
Shanxi	5.8	5.6	–0.2	–3.4	0	170	25.0	29.4	4.4	5.9	1	1.7	4.4
Neimenggu	18.0	17.2	–0.8	–4.4	0	720	6.7	8.1	1.4	7.0	1	1	17.8
Liaoning	14.1	13.6	–0.5	–3.5	0	320	24.8	34.4	9.6	12.9	1	1.3	4.1
Jilin	10.7	11.3	0.6	5.6	+0.8	420	14.2	16.1	1.9	4.5	1	2	9.3
HeilongJiang	18.2	29.6	11.4	62.6	+8.9	800	8.7	11.0	2.3	8.8	1	5.9	11.5
Shanghai	10.8	10.8	0	0	0	650	30.7	41.6	10.9	11.8	1	5.5	1.1
Jiangsu	50.3	44.5	–5.8	–11.5	0	600	12.9	19.2	6.3	16.3	1	3.4	6.3
Zhejiang	20.2	20.1	–0.1	–0.5	0	430	24.7	30.3	5.6	7.6	1	1.9	4.4
Anhui	16.3	17.6	1.3	8.0	+1.1	300	14.6	17.2	2.6	5.9	1	2.3	7.2
Fujian	15.6	17.6	2.0	12.8	+1.8	510	17.4	22.2	4.8	9.2	1	2.4	5.8
Jiangxi	23.2	21.7	–1.5	–6.5	0	530	7.1	9.2	2.1	9.9	1	2.7	8.8
Shandong	25.0	24.3	–0.7	–2.8	0	270	24.0	34.4	10.4	14.4	1	1.8	7.2
Henan	20.4	20.4	0	0	0	220	16.9	25.0	8.1	16.0	1	1.4	4.3
Hubei	23.8	27.0	3.2	13.4	+1.9	450	15.4	15.9	0.5	1.1	1	2.8	6
Hunan	32.0	31.5	–0.5	–1.6	0	490	9.4	11.6	2.2	7.8	1	1.4	5.8

TABLE 3 Water Use Indices for Administrative Regions in China (continued)

Item / Administrative Region	Total Water Use (10^9 m^3)		Increase of Water Use			Per Capita Water Use (2000) (m^3)	Water Use Efficiency (yuan GDP/m^3)				Water Use Structure (2000)		
	1993	2000	Increment	%	Rate of Increase		1997	2000	Increment	Rate of increase	Water supply	Industry	Agriculture
Guangdong	41.7	42.9	1.2	2.9	+0.4	500	15.2	22.2	7.0	15.4	1	1.5	3.8
Guangxi	21.5	29.2	7.7	35.8	+5.1	650	7.2	6.9	–0.3	–1.5	1	1.3	7.7
Hainan	4.2	4.4	0.2	4.8	–0.7	560	10.0	11.7	1.7	5.7	1	0.8	7.2
Chongqing	27.6	24.5	–2.5	–10.7	+1.6	180	24.0	28.6	4.6	6.4	1	2	1.5
Sichuan	23.3	20.8	–2.5	–10.7	–1.5	250	16.4	19.2	2.8	5.7	1	1.8	5
Guizhou	5.7	8.4	2.4	42.1	+6.0	240	10.1	11.8	1.7	5.6	1	1.3	3.3
Yunnan	10.8	14.7	3.9	36.1	+5.2	340	12.0	13.3	1.3	3.6	1	1.1	6.6
Tibet	1.6	2.7	1.1	68.8	+9.8	1040		4.3			1	0.4	13.8
Shanxi	8.2	7.8	–0.4	–4.9	0	220	16.5	21.3	4.8	9.7	1	1.2	5.5
Gansu	11.8	12.2	0.4	3.4	+0.5	480	6.6	8.0	1.4	7.1	1	2.3	12.8
Qinghai	2.7	2.7	0	0	0	540	7.6	9.4	1.8	7.9	1	1.3	1.5
Ningxia	8.6	8.7	0.1	1.2	+0.2	1550	2.0	3.0	1.0	16.7	1	2.8	47.5
Xinjiang	48.2	47.9	–0.3	–0.6	0	2490	2.4	2.8	0.4	5.6	1	0.7	28.6

Source: Ke Li Dan (2002).

B. Significant Structural Change of Water Use

The ratio of water use in the agricultural sector (including rural water supply) to total water use dropped from 97.1% in 1949 to 74.1% in 2000, while that of industrial and urban water supply increased from 2.3 to 20.7% and from 0.6 to 5.2%, respectively. The portion used for irrigation in the agricultural sector went from 91% in 1980 to 85% in 2000. In this 20-year period, irrigated land increased to 6.7 million ha (a one-eighth increase compared to the pre-1980 period). However, water use for irrigation remained at about 350 billion m^3 each year, which means that during this period it remained nearly unchanged. Furthermore, water use in the industrial sector has remained at about 110 billion m^3 since 1997. Figure 1 shows the structural change of water use in China.

C. Steady Decline of Per Capita Water Use After 1980

Per capita water use is an integrated index reflecting the level of water use for all user sectors. The per capita water use increased from 187 m^3 in 1949 to 452 m^3 in 1980 and steadily declined to 430 m^3 in 2000. The per capita water use also varies greatly with different administrative regions. Shanxi

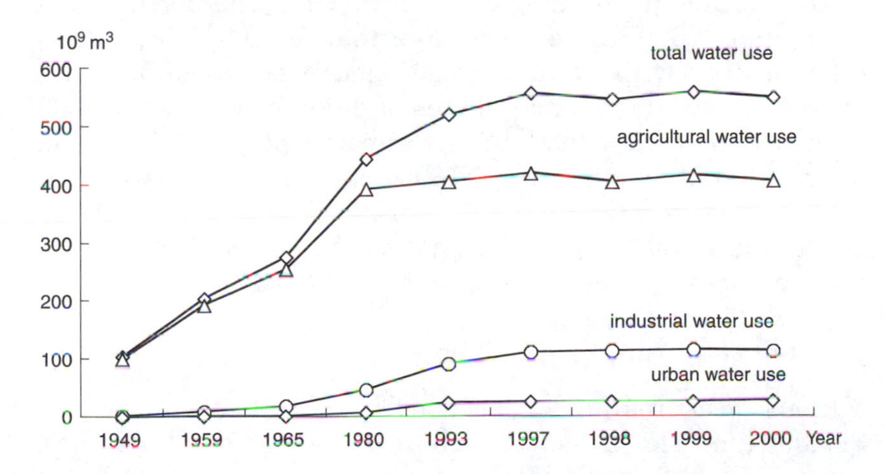

Figure 1 Water use for different sectors (1949–2000). (From Ke Li Dan, 2002.)

Province ranks the lowest (187 m^3) whereas the Xinjiang Autonomous Region ranks the highest (2490 m^3), more than 13 times higher than Shanxi. This is due to differences in climate and physiographic conditions and to traditional ways of water use. The index remained stable for different administrative regions. The per capita water use varies with different regions, but the figures of per capita water use for each of the different regions remains stable for a period of time.

IV. DROUGHT CONTROL AND MANAGEMENT

To cope with drought in the agricultural sector, China has implemented an integrated approach that consists of engineering, legal, institutional, economic, and technical measures.

A. Building a Drought Control and Management System

1. Developing Irrigation Systems

In the early 1950s, efforts concentrated on restoration, expansion, and renovation of old irrigation systems. Storage facilities, including ponds and reservoirs of different sizes, have developed rapidly since the 1960s. In the 1970s, to meet the increasing demand of food supply, irrigation from canals was promoted in the south and well irrigation prevailed in the north China plain. As of 2000, there were more than 85,000 reservoirs of different sizes with a total storage capacity of 518.4 billion m^3, more than 30,000 diversion gates of different types, and 3.99 million wells with a total storage capacity of 35.9 million kw. Irrigated land accounts for 42% of total cultivated land (130 million ha) and provides more than two-thirds of the national grain output, 80% of the national cotton output, and 90% of the national vegetable crop (Zhang Shi Fa, 2002).

2. Soil and Water Conservation

Various measures have been used to control soil erosion on 809,000 km^2. The measures include terracing (9.6 million ha), construction of check-dams in gullies and small creeks to create usable land (1.95 million ha), afforestation (39.9 million ha), and restoration of grassland (4.3 million ha).

3. Developing Rain-Fed Agriculture

Because more than 50% of the cultivated land area is rain fed, farming has been adapted to make full use of rainwater. Adaptations include deep plowing and soil loosening to increase the storage capacity of the top soil and developing drought-resistant crops and associated cultivation techniques. Since the 1980s, many pilot projects based on indigenous techniques have begun in arid and semiarid regions. In addition, small rainwater harvesting facilities are also being developed in the north and northwest.

4. Drought Monitoring, Planning, and Management during Dry Periods

China's comprehensive hydrological network has played an important role in flood and drought monitoring. The real-time monitoring of the drought process is handled primarily by provincial governments, and the central government closely monitors the development of drought and gives guidance and funding support to local governments to combat drought. Long-term meteorological forecasting is the responsibility of the Meteorological Department in coordination with the Water Department, but it has not played a significant part in drought management.

Drought management has followed the traditional approach, which largely relies on crisis management. This reactive approach has been ineffective in responding to drought, and the government has recently initiated a nationwide project, Strategy on Drought Management, to cope with the problem. A number of pilot projects on risk management are being implemented in several provinces.

B. Improving Water Use Efficiency

1. Improving Irrigation Technique

The agricultural sector is the main user of water, but water use efficiency is only 45% for the entire irrigation system. The losses are mainly due to seepage in the irrigation system as well as the use of flood irrigation in many places, especially in southern China, where farmers used to follow traditional technology in applying water to the rice field. Water-saving techniques have been promoted in grain-producing areas, particularly in arid regions. In 1997, China implemented a water-saving program on 15.3 million ha of irrigated land using different techniques, such as "anti-seepage" measures for canals (8.7 million ha), water conveyance with pipes (5.2 million ha), and sprinkler irrigation (1.2 million ha).

2. Promoting a Water Tariff System

A water tariff system for irrigation was initiated and has been promoted since the 1980s. It has played an important role in addressing the challenge of water shortage in China. The problem is that the fee is too low for most areas, which means less incentive to the farmers to use water efficiently. The water tariff is being adjusted so that it will at least meet the operation requirements and maintenance costs of irrigation systems. In 1996, the central government decided to select 300 counties from more than 2000 as pilot counties to implement the water-saving program.

C. Institutional Building for Drought Management

1. Establishing National Flood Fighting and Drought Relief System

Flood Fighting and Drought Relief Headquarters (FFDRHQ) were established in 1992 for the government at the county level and above. Members of the FFDRHQ include all the relevant departments at the same level led by the head of local governments. The central government is headed by a vice premier designated by the state council (cabinet) with the executive office affiliated with the Ministry of Water Resources. The major responsibilities of FFDRHQ for drought relief cover the following (Department of Policy and Regulation, Ministry of Water Resources, 1998):

- Operating the drought monitoring, forecasting, and warning system in coordination with the Meteorological Department and reporting drought information to the government and public

- Organizing the preparation of a drought relief scheme and supervising its implementation
- Coordinating activities of all the parties concerned with drought response during the dry period
- Allocating the funds from the central government to the drought-affected areas
- Preparing and supplying materials for drought relief
- Transferring inter-basin water and allocating real-time water for important rivers (cross-boundary administration)

The Department of Water Resources at all levels handles most of the government functions in drought relief (policy guidance; resource mobilization, including financial and human resources; and servicing, including monitoring, reporting, warning, technical advice, and post-relief). The members include the State Development and Reform Commission, State Economic and Trade Commission, Ministry of Public Security, Ministry of Interior, Ministry of Finance, Ministry of Land Resources, Ministry of Construction, Ministry of Railway, Ministry of Communication, Ministry of Information Industry, Ministry of Agriculture, National Bureau of Aviation, National Bureau of Broadcasting and Television, National Bureau of Meteorology.

The FFDRHQ was established to reinforce leadership and inter-agency coordination. Members at the central level assume their respective responsibilities as assigned by the state council; the responsibilities of the local-level members of FFDRHQ are similar to those of the central government members.

2. Organizing Drought Relief Teams in Rural Areas

Since the 1960s, drought relief teams have been organized in many provinces and regions. The teams are nongovernmental, nonprofit, and professional organizations at the grassroots level, providing technical support to farmers. Both the government and the beneficiaries provide funding. As of late 2003, more than 10,000 teams were organized and equipped with 210,000 irrigation machineries of different types. This institutional setup has led to efficient use of resources because it concentrates the government's input into the teams instead of directly to the drought-affected farmers. Drought relief teams are being built in rural areas for all the provinces and autonomous regions of the country.

D. Legal, Regulatory, and Policy Measures

In the past 20 years, the government undertook several legal, regulatory, and policy measures to improve the water use efficiency of different users. The momentum of economic reform has greatly promoted the restructuring of water user sectors and efficient use of water.

1. Water Law

The Water Law of the People's Republic of China, enacted in 1998, provided the legal basis for water resources planning, development, and management. The revised Water Law was promulgated in August 2002. It keeps the main structure of the Water Law and amends, complements, and stresses some important provisions:

- Emphasizes integrated water resources planning and stresses that water resources planning should be an integral part of the national economic plan
- Emphasizes environmental protection in the development of water resources, including water quality control and water source protection for all natural water bodies
- Highlights rational water allocation and water saving
- Adds unified management of a river basin so that the legal position of river basin organizations is identified
- Clarifies treatment of water disputes to include consultation, mediation, and lawsuits

2. Water-Withdrawing Permit System

A water-withdrawing permit system has been built and applied to all projects directly withdrawing water from rivers, lakes, and underground. More than 60,000 licenses have been issued, totaling 400 billion m^3 of water withdrawal.

3. Water-Saving Policy

Water saving has become national policy through various measures, including improvement of water use management and innovations in water-saving techniques to reduce water consumption and increase the rate of reuse in the industrial sector. Areas short of water are restricted from developing industry and agriculture that require massive water use. Water use is metered in urban areas; for industry, a progressive water tariff has replaced a system of fixed fees.

4. System Reform of Water Charge and Levy of Water Resources Fee

In 1985, the state council issued a regulation on the levy of a water charge, thus leading to the end of a history of free water use. The water resources fee is levied for those who withdraw directly from both underground and surface water bodies.

V. PERSPECTIVES ON WATER SUPPLY AND DEMAND IN THE 21ST CENTURY

A. Low Level of Per Capita Water Use

The United Nations (U.N.) categorizes China as a water-vulnerable country, along with the United States, Japan, and many European countries (UNDDSMS, 1996). North China, specifically the Hai, Huai, and Yellow rivers, covers 1.44 million km^2 (15%) of China's territory and accounts for 39.4% of total cultivated land, 34.7% of the population, and 32.4% the country's GDP. The per capita water availability for the three basins is less than 500 m^3 (Liu Chang Ming and Chen Zhi

Kai, 2001) and the percentage of withdrawal is much higher (>50%). China must maintain low levels of water use to support the process of sustainable development. China cannot reach the same level of per capita water use as the United States, Japan, and other European countries in view of its huge population and its extremely uneven areal distribution of water resources.

B. Anticipation of Water Use Trend

Table 2 shows that from 1949 to 1980, water use quadrupled. During that time, the population nearly doubled (1.78) and the total grain output tripled. In the transitional period from 1980 to the present, China entered an era of economic reform, and the water use trend grew slowly to sustain the major change that the country's economy was experiencing. This resulted from structural changes in water use and improvements in water use efficiency. Compared to other developed countries, China has considerable potential to raise water use efficiency. The integrated index of water use efficiency in China was US$2/m^3 in 2000 (US$13.1 for the United States in 1995). Water use for irrigation has remained unchanged (350 billion m^3) for more than 20 years since 1980, and water use efficiency currently is only 0.45, so there is much room for improvement.

The structural change of water use may also accelerate. Under the process of structure reform and the development of a market-oriented economy, water use tends to meet the needs of those sectors that bring more economic benefit, so the gradual reduction of the proportion of agricultural water use to total water use is acceptable. Meanwhile, the application of new technologies and the blooming of tertiary industry will make water use reduction more attractive to industry. Based on the experiences of big cities such as Beijing, Shanghai, and Tianjin, the rapid expansion of urban areas and the development of the economy did not lead to an increase of per capita water use in these cities. It can be anticipated that per capita water use may not increase as the momentum of restructuring continues.

It should be pointed out that overestimates of water demand in the past exaggerated the estimated gap between water supply and demand. In the early 1980s, the estimated water demand for China in 2000 was projected to be 700 billion m^3, on the basis of a nationwide water use project (Department of Planning, Ministry of Water Resources, 1987). The government made similar estimates about water demand for the entire country and for different administrative regions. Overestimation of water demand led to mistakes in strategy and policy formulation. The international concern in the 1990s about the impact of large grain imports from China on world food security perhaps also originated from the overestimate of water demand.

VI. CONCLUSION

Reviewing water use of the past 20 years, one can conclude that the low increase (1%) in the water use rate supported the rapid economic growth at the average rate of 9%; the zero growth of irrigation water lasted for 20 consecutive years while the grain output increased 44.3% under the expansion of 6.7 million ha of irrigated land. The steady decline of per capita water use and the zero growth of water use for many administrative regions in recent years indicate that the gap between water supply and demand can be bridged, if useful experiences and lessons are appropriately summed up and learned.

The low or zero growth of water use in water-scarce regions, especially in the north China plain, was achieved at the cost of environmental degradation. Improvement of the water environment should be an inseparable part of water management in the course of sustainable development in the 21st century.

Recommended drought management actions (and their current status) include:

1. Formulation of drought relief plans of different frequencies on the basis of the water resources plan. The first phase of this recommendation has started.

2. Increased financial support for drought relief. The program was started 3 years ago, but support has been inadequate.
3. Improvements to the management of drought relief. This recommendation mainly refers to the reform of irrigation projects, such as changing the ownership of small projects, and promoting a contract system with users (farmers) for sustainable development.
4. Guidelines to identify and assess drought disasters on a scientific basis.
5. Legislation to promote the system of drought preparedness. We expect drought management regulations will be formulated and approved in the near future.

REFERENCES

Department of Hydrology, Ministry of Water Resources. *Water Resources Assessment for China*. Beijing: China Water and Power Press, 1992.

Department of Planning, Ministry of Water Resources. *Water Use in China* (in Chinese). Beijing: China Water and Power Press, 1987.

Department of Policy and Regulation, Ministry of Water Resources. *Roles and Functions of the Ministry of Water Resources and its Relevant Departments* (in Chinese). Beijing. 1998.

Ke Li Dan. "An analysis of trends on water use in China in 1949–2000—Comment on the starting point of south–north water transfer project" (in Chinese). *Science and Technology Review* 9–10:12, 2002.

Liu Chang Ming, Chen Zhi Kai. *The Assessment of China's Water Resources and the Trend of Water Supply and Demand* (in Chinese) (pp. 3–4). Beijing: China Water and Power Press, 2001.

UNDDSMS. 1996. Information sheets on classification of countries by water availability and use and other socioeconomic indices (in Chinese).

UNESCO-WMO. Hydrological aspects for droughts. *Studies and Reports in Hydrology* 39, 1985.

Zhang Hai Lun. *Introduction in Flood and Drought Disaster in China* (in Chinese). Beijing: China Water and Power Press, 1997.

Zhang Shi Fa. "Study on Flood and Drought Mitigation System. Nanjing: Nanjing Research Institute of Hydrology and Water Resources" (in Chinese), 1999.

Zhang Shi Fa. Flood control, drought prevention, disaster reduction. In: Zhou Er Ming, Zhao Guang He. *Strategic Study of China's Water Resources Development* (in Chinese) (pp. 37–38). Beijing: China Water and Power Press, 2002.

13

A Role for Streamflow Forecasting in Managing Risk Associated with Drought and Other Water Crises

SUSAN CUDDY, REBECCA LETCHER, FRANCIS H. S. CHIEW, BLAIR E. NANCARROW, AND TONY JAKEMAN

CONTENTS

I. INTRODUCTION

Climatic variability should be a significant factor influencing agricultural production decisions. Historically in Australia, farmers and governments have invested heavily in reducing the influence of this variability on agricultural production. This investment has included construction of large dams on major river systems throughout the country, primarily for irrigation purposes, and allocation and development of groundwater resources. This development policy placed large pressures on ecosystems and has significantly modified river systems. In 1994, the Council of Australian Governments began a period of water reform, entering a new management phase for water resources. These reforms have included assessment of the sustainable yield from aquifer systems, often found to be below current allocation and even extraction levels, as well as allocation of a proportion of flows to the environment. In many catchments these water reforms have not only reduced irrigators' access to some types of water but have also implicitly increased the effect of climate variability on their decision making by increasing their reliance on pumping variable river flows.

These management and allocation pressures are compounded by Australian streamflow (and to a lesser extent climate) being much more variable than elsewhere in the world. The interannual variability of river flows in temperate Australia (and southern Africa) is about twice that of river flows elsewhere in the world (Figure 1; Peel et al., 2001). This means that temperate Australia is more vulnerable than other countries to river flow–related droughts and floods. In such a challenging environment, forecasting tools that support improved decision making resulting in efficiencies in water use and reduced risk taking are highly desirable. The development and use of such tools is the focus of considerable research and extension activity in government and industry.

A. Seasonal Forecast and Climate Variability

Relationships between sea surface temperatures and climate are well documented. The relationship between Australia's hydroclimate and the El Niño/Southern Oscillation (ENSO)

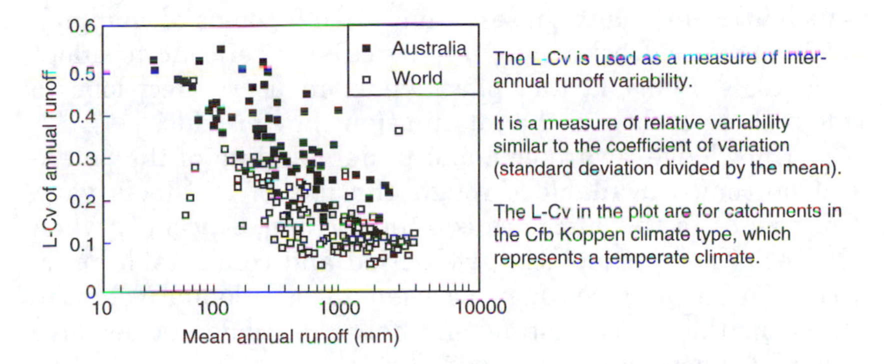

Figure 1 Interannual variability of Australian streamflow relative to the rest of the world.

is among the strongest in the world (Chiew and McMahon, 2002). El Niño describes the warm phase of a naturally occurring sea surface temperature oscillation in the tropical Pacific Ocean. Southern oscillation refers to a seesaw shift in surface air pressure at Darwin, Australia, and the South Pacific island of Tahiti. Several indices have been derived from this relationship, in particular the Southern Oscillation Index (SOI), which describes the Tahiti minus Darwin sea level pressure and is commonly used as an indicator of ENSO. The strong relationships that exist between climate, streamflow, and ENSO form the scientific basis for forecast tools developed throughout Australia and other parts of the world. In the Australian context, the Bureau of Meteorology routinely provides seasonal climate outlooks (e.g., probability that the total rainfall over the next 3 months will exceed the median) and computer packages such as *Rainman Streamflow* (Clewett et al., 2003) are heavily promoted. The patchy adoption of these tools, and thus the inability to reap the perceived gains in water use efficiency, is of concern to their promoters and research and development agencies.

B. Adoption Constraints

A major issue for the designers of decision support tools is the degree of likely uptake by the potential users, and this is no different for seasonal forecasting. The farming community, which is traditionally conservative when it comes to changing well-entrenched behaviors, is particularly reticent to adopt such tools. Many factors play a part in users' decisions to adopt these tools and the information they provide.

Knowledge, awareness, and understanding of the potential outcomes available through the use of the tools vary. Confidence in the outcomes is often lacking, especially when the tools may be replacing well-tried and comfortable practices. These practices may be seen to be adequate for the decisions they are assisting, and hence users do not perceive a need for new technologies.

Previous experiences associated with the technologies being used by the tools will also be a factor. These may be first-hand experiences or purely word of mouth in the community. Local opinion will frequently be more powerful than information from "outsiders." Naturally, if past experiences have resulted in negative consequences, the uptake of the new technology will be even less likely. Confidence in the new technology and trust in the provider of the technology are therefore likely to be highly influential. In fact, the "human factor" frequently can be less certain than the technologies themselves.

II. ESTIMATING THE POTENTIAL

Most investigations of the potential of forecast tools compare their predictions against "no knowledge." This section describes the coupling of forecast models to models that simulate a range of water management behaviors within a constrained problem definition. Quantification of the net financial return to irrigators of adopting climate forecasts as part of their decision-making process would provide a strong measure of the benefit of these forecasts. This is tempered by an analysis of the potential market, which reveals that a significant improvement in reliability and relevance is required before widespread adoption can be considered.

A. Case Study Context

To consider the potential benefits to agricultural production of seasonal forecasts, we investigated their potential impact on farm-level decisions and returns in an irrigated cropping system. We premised that the potential benefit of seasonal forecasts was probably greatest in a farming system subject to significant uncertainty. For this reason, the farming system represented in the decision-making models is that of an irrigated cotton producer operating on an unregulated river system, relying on pumping variable river flows for irrigation purposes during the season. This type of farm is typical in unregulated areas of the Namoi basin in the northern Murray-Darling basin, particularly the Cox's Creek area (Figure 2). However, for this analysis, the modeling should be considered to represent a theoretical or model farm rather than a farm from a particular system; the value of forecasts was tested on this farm using forecasts and flows from many

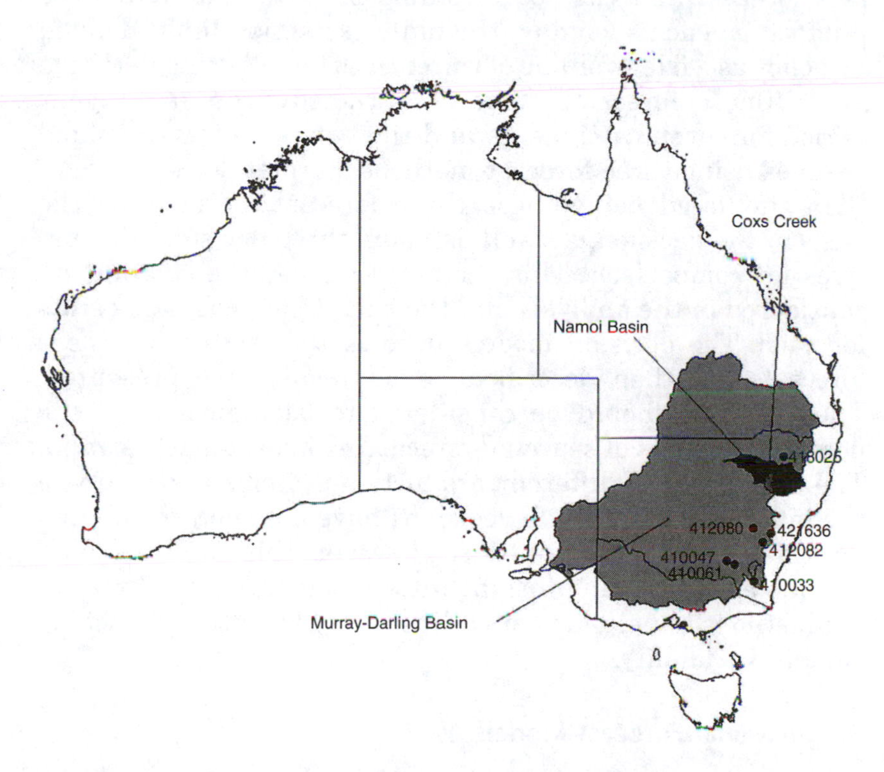

Figure 2 Map of the case study area highlighting the Cox's Creek region of the Namoi basin within the Murray-Darling basin system of eastern Australia.

different river systems in eastern New South Wales. We did this to test the sensitivity of the results and recommendations to the hydrology and climate of the river system.

Given that the model farm is assumed to be pumping from the river for irrigation supply, production and water availability are limited by the number of days on which the farm can pump flows from the river. To mimic the types of flow rules on these unregulated systems and to test the sensitivity of results to these rules, two pumping thresholds were considered—the 20th and 50th percentile of flow (i.e., flow that is exceeded 20% or 50% of the time).

The forecast provided for each year is the number of days that are above these pumping thresholds (i.e., the number of days on which pumping is allowed). The model farmer factors this forecast and the total volume of water allowed to be pumped on each such day (the daily extraction limit, defined by policy as a fixed volume of water) into the planting decision.

Climate forecasts were constructed over an 86-year period for seven catchments and the two pumping threshold regimes using three forecast methods. Farmer decisions were then simulated using these three forecast methods as the basis of the decision, as well as using three decision alternatives for comparison. This section describes the catchments considered in the analysis and the climate forecasting results for each. The decision models used in the analysis of these forecasts are then described before results are presented. These results should be considered to be indicative of the potential benefits of seasonal forecasting in eastern Australia. The complexity of different production systems and many of the influences on real-life decisions have not been considered for this preliminary analysis. However, this analysis does provide an interesting insight into the potential for forecasting methods to help farmers adjust away from the impacts of climate variability.

B. Seasonal Forecast Models

The relationship between streamflow and ENSO and the serial correlation in streamflow can be exploited to forecast streamflow several months ahead. These relationships are well described in Chiew and McMahon (2003) and demonstrate the statistical significance of the lag correlation of the linear relationship between 3-month streamflow (in Oct–Nov–Dec and in Jan–Feb–Mar) and the SOI value in the previous 3 months in catchments throughout Australia. Using this relationship, we can forecast summer streamflow throughout most of eastern Australia from spring indicators of ENSO. Serial correlation in streamflow must also be considered when forecasting streamflow because it is generally stronger than the streamflow–ENSO relationship and is persistent throughout the year.

To make risk-based management decisions, we must express forecasts as exceedance probabilities (e.g., probability of getting at least 10 pumping days). In this study, exceedance probability forecasts are derived at tributary scale for seven unimpaired catchments in the Murray-Darling basin. The derivation of the forecasts is based on categorization and consequent nonparametric modeling of streamflow distributions and their antecedent conditions (e.g., discrete SOI categories) (see, for example, Sharma, 2000, and Piechota et al., 2001). Catchments were selected because of their relative proximity to the Namoi basin (all within the Murray-Darling basin in New South Wales) and to reflect a range of rainfall–runoff conditions and forecast skills. Proximity to the Namoi basin is to support coupling with the decision-making models that have been developed by Letcher (2002) within the water management regulatory framework in the Namoi basin, although they simulate representative farmer behavior.

Daily streamflow data from the period 1912–1997 are used. The data include extended streamflow estimates using a conceptual daily rainfall–runoff model (Chiew et al., 2002). The catchment locations and long-term average rainfall–runoff characteristics are summarized in Table 1.

Forecasts are made for the number of days in October–February that the daily flow exceeds the two pumping thresholds under consideration. The thresholds are calculated based on flow days only, defined as days when the daily flow exceeds 0.1 mm. The forecast is derived by relating the number of days in October–February that the daily flow exceeds a threshold to explanatory variables available at the end of

TABLE 1 Summary of Characteristics for Catchments Used in the Analysis

Catchment	Catchment and Rainfall–Runoff Characteristics								
	Lat.	Long.	Area (km^2)	Rainfall (mm)	Runoff (mm)	Runoff Coef. (%)	% Days Flow >0.1 mm	Percentile Flows (mm) 20%	Percentile Flows (mm) 50%
410033 Murrumbidgee R @ Mittagang Crossing	36.17	149.09	1891	882	134	10–15	71	0.55	0.28
410047 Tarcutta Ck @ Old Borambola	35.15	147.66	1660	818	110	10–15	50	0.68	0.31
410061 Adelong Ck @ Batlow Road	35.33	148.07	155	1138	256	>20	89	0.97	0.44
412080 Flyers Creek @ Beneree	33.50	149.04	98	915	106	10–15	50	0.65	0.29
412082 Phils Creek @ Fullerton	34.23	149.55	106	821	124	10–15	62	0.58	0.27
418025 Halls Creek @ Bingara	29.91	150.58	156	755	44	6	24	0.22	0.14
421036 Duckmaloi River @ Below Dam Site	33.77	149.94	112	967	244	>20	80	0.95	0.40

September. The explanatory variables used are the SOI value averaged over August and September and the total flow volume in August and September. We derive the forecast using the nonparametric seasonal forecast model described in Piechota et al. (2001) and express it as exceedance probabilities. Such forecasts closely approximate low-risk decision-making behavior and can be used as a direct input into the decision-making models.

Three forecast models are used:

1. FLOW: Forecast derived from flow volume in August–September
2. SOI: Forecast derived from SOI value in August–September
3. FLOW+SOI: Forecast derived from flow volume and SOI value in August–September

C. Forecast Model Results

All models exhibit significant skill in the forecast, summarized in Table 2. Two measures of forecast skill are used—Nash-Sutcliffe E and LEPS scores.

The Nash-Sutcliffe E (Nash and Sutcliffe, 1970) provides a measure of the agreement between the "mean" forecast (close to the 50% exceedance probability forecast) and the actual number of days in October–February that the daily flow exceeds a threshold. A higher E value indicates a better agreement between the forecast and actual values, with an E value of 1.0 indicating that all the "mean" forecasts for all years are exactly the same as actual values.

The LEPS score (Piechota et al., 2001) attempts to compare the distribution of forecast (forecast for various exceedance probabilities) with the number of days in October–February that the daily flow exceeds a threshold. A LEPS score of 10% generally indicates that the forecast skill is statistically significant. A forecast based solely on climatology (same forecast for every year based on the historical data) has a LEPS score of 0.

TABLE 2 Summary of Forecast Skills for Catchments Used in the Analysis

Catchment	Case	FLOW		SOI		FLOW+SOI	
		E	LEPS	E	LEPS	E	LEPS
410033 Murrumbidgee R @ Mittagang Crossing	Days >20%	0.35	27.1	0.23	11.6	0.58	41.7
	Days >50%	0.36	23.1	0.19	12.2	0.60	39.7
410047 Tarcutta Ck @ Old Borambola	Days >20%	0.41	32.8	0.23	17.6	0.57	46.4
	Days >50	0.39	26.2	0.18	11.2	0.50	36.0
410061 Adelong Ck @ Batlow Road	Days >10%	0.54	41.4	0.16	12.0	0.64	49.5
	Days >20%	0.63	42.0	0.17	11.1	0.71	50.4
412080 Flyers Creek @ Beneree	Days >20%	0.34	25.8	0.22	10.2	0.54	37.6
	Days >50%	0.42	28.8	0.22	10.9	0.56	40.0
412082 Phils Creek @ Fullerton	Days >20%	0.40	19.2	0.22	12.3	0.59	32.1
	Days >50%	0.54	30.0	0.22	12.2	0.64	39.7
418025 Halls Creek @ Bingara	Days >20%	0.13	12.4	0.16	11.7	0.29	26.3
	Days >50%	0.26	15.3	0.16	13.0	0.44	31.5
421036 Duckmaloi River @ Below Dam Site	Days >20%	0.16	12.3	0.24	13.5	0.45	28.1
	Days >50%	0.24	16.7	0.27	17.7	0.51	34.0

The LEPS scores in all the forecast models are greater than 10%, indicating significant skill in the forecast. The SOI model has similar skill in the seven catchments, with E values of about 0.2 and LEPS scores of 10–15%. The FLOW model is considerably better than the SOI model in five catchments (410033, 410047, 410061, 412080, 412082; E generally greater than 0.35 and LEPS generally greater than 25%), whereas at the gauge sites of the other two catchments (418025, 421036), the FLOW and SOI models have similar skill. In all seven catchments, the FLOW+SOI model has greater skill than the FLOW or SOI model alone. In the five catchments where the FLOW model has greater skill than the SOI model, the E and LEPS for the FLOW+SOI model are generally greater than 0.5 and 40%, respectively (compared to 0.35 and 25% in the FLOW model). In the two catchments where the FLOW model and SOI model have similar skill, the E and LEPS for the FLOW+SOI model are generally greater than 0.3 and 25% (compared to less than 0.25 and 20% in the FLOW or SOI model alone).

D. Decision-Making Models

All decisions were modeled using a simple farm model that assumed that farmers act to maximize gross margin each year, given constraints on land and water available to them in the year. This model is a modified version of a decision model for the Cox's Creek catchment developed by Letcher (2002). Total farm gross margin was analyzed for all catchments, pumping thresholds, and forecast methods using four possible decision methods:

1. *Seasonal forecast decision*. The decision is made assuming that the 20th and 50th percentile exceedance probability forecasts (using SOI, FLOW, and SOI+FLOW) for the number of pumping days are correct.
2. *Naïve decision*. The decision is made assuming that the number of pumping days this year is equal to the number of pumping days observed last year.

3. *Average climate decision*. The decision is made assuming that the number of days for which pumping is possible in each year is the same and equal to the average number of days pumping is permitted over the entire 86-year period.
4. *Perfect knowledge decision*. The decision is made with full knowledge of the actual number of days on which pumping is possible in each year. This is essentially used to standardize the results, because it is a measure of the greatest gross margin possible in each year given resource constraints.

The same simple farm model is used in all cases. This model allows the farm to choose among three cropping regimes—irrigated cotton with winter wheat rotation, dryland sorghum and winter wheat rotation, and dryland cotton and winter wheat rotation. Production costs are incurred on crop planting, so areas planted for which insufficient water is available over the year generate a loss. For such crops, it is assumed that the area irrigated is cut back and a dryland yield is achieved on the remaining area planted.

E. Modeling Results

We ran models for each catchment over the 86-year period for every combination of pumping threshold, forecast, and decision-making method. The total gross margin achieved by the farm over the entire simulation period under each of the decision models and forecasting methods is charted for the 20th percentile (Figure 3) and the 50th percentile (Figure 4) pumping threshold, respectively. In each figure, the x axis labels the seven catchment identifiers and the y axis is the total gross margin in Australian dollars.

These figures lead to a consistent set of observations:

- Use of any of the three forecast methods leads to a greater gross margin than either the average or the naïve decision methods.
- In general, the SOI+FLOW method gives the greatest gross margin of the three forecast methods, with SOI generally providing the lowest gross margin.

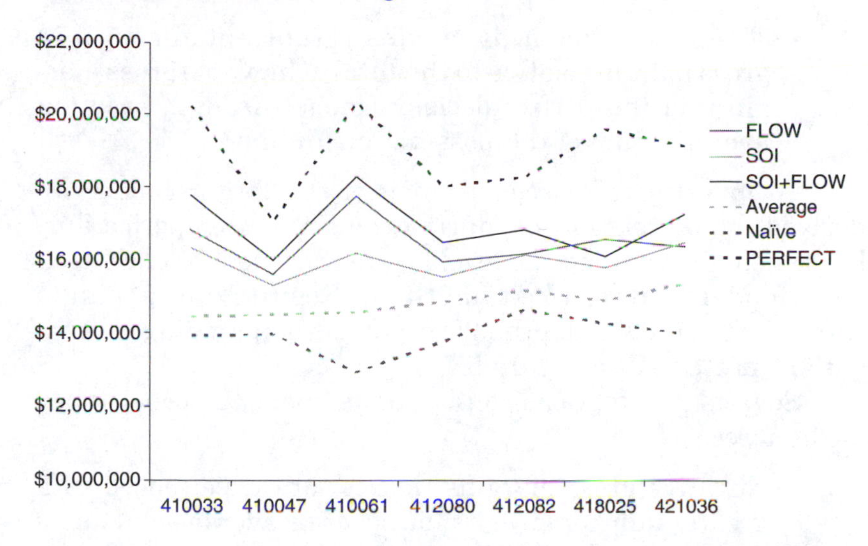

Figure 3 Total profit (annual gross margin) over 86-year simulation period for each catchment using different decision methods for pump threshold at the 20th percentile of flow.

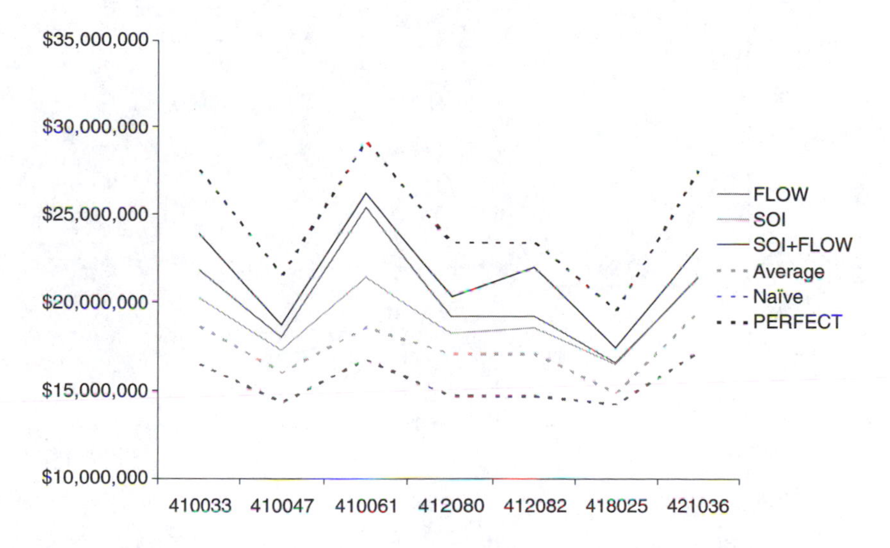

Figure 4 Total profit (annual gross margin) over 86-year simulation period for each catchment using different decision methods for pump threshold at the 50th percentile of flow.

- The forecast methods provide a substantial return in gross margin relative to the total achievable gross margin (via the perfect decision model) in each case (on average, 55% of the possible maximum).

To investigate the consistency of the forecast skill, we derived the percent of time during the simulation period during which different income levels were exceeded for each decision model and forecast method. Results for a single catchment (410033) and the 20th percentile pumping threshold are presented in Figure 5.

Several observations can be made about the consistency of the forecasts:

- Negative gross margins (losses) are experienced in a greater number of years for both the average and naïve decision methods (>7% of time) than for any of the seasonal forecast methods (<3.5%).

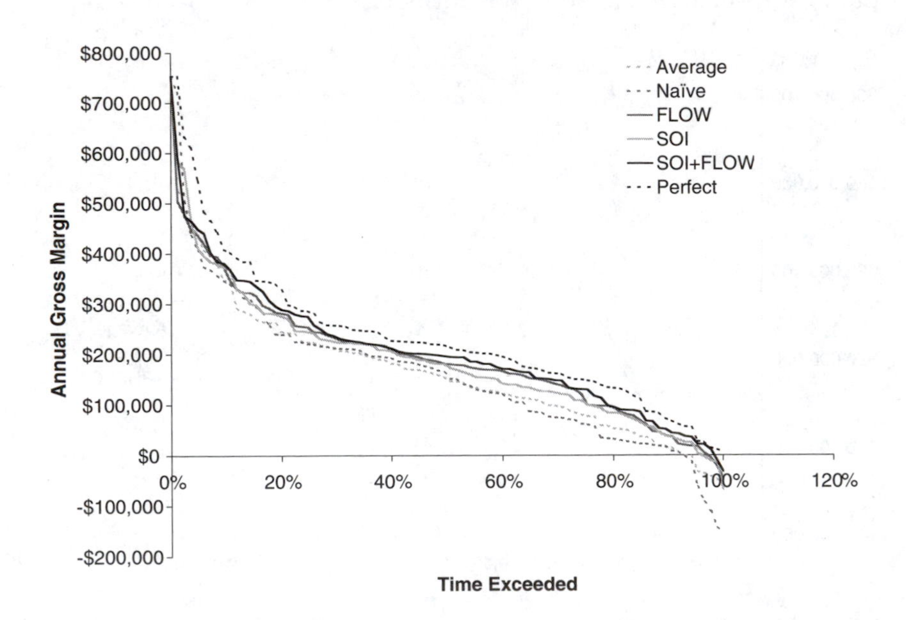

Figure 5 Exceedance probability for annual gross margin for one catchment (410033) with a pumping threshold at the 20th percentile of flow.

- The naïve and average decision methods give a lower income at almost all exceedance probabilities, and for those areas where they are greater, the difference is very small.
- The naïve decision method gives a greater gross margin for very high gross margin years (2.4% of the time).

III. REALITY BITES

The integrated modeling approach developed for this study demonstrates the gains that can be made by the routine incorporation of seasonal forecasting into water management decision.

To test the likeliness of farmers to use the seasonal forecasting tools, we conducted a series of ten semistructured, scoping interviews. The interviewees were irrigators on a nonregulated tributary of the Namoi River (in the Upper Murray-Darling basin) and were therefore highly dependent on the river flows. With the uncertainty of annual water supplies to irrigate their crops, it was thought that this group might be more positive about seasonal forecasting than those on regulated rivers.

Given the small sample number, we compared the data collected with that from a similar study conducted by others in the southern Murray-Darling basin (URS Australia, 2001). That study consisted of 29 interviews followed by a workshop with six participants. The findings of both studies were similar, thus providing confidence in the outcomes of these limited interviews.

Knowledge and understanding of the term and the forms of seasonal forecasting were highly variable, ranging from a good understanding to little or misguided awareness. Participants seemed to misunderstand the difference between types of forecasting and sources of forecasting. However, considerable support exists for natural signals rather than the use of technology—for example:

> Some of the best indicators in times of drought have been the ants' nests around the house—if there is a lot of movement by the ants, it's generally going to rain soon.

> There are many natural signs that are more useful than the scientific information we are given.

The degree to which people understood probabilities, or thought they could be useful to them, was also variable. Many were skeptical of the probabilities given their derivation by the extrapolation of past data to the present. Usefulness was also questioned in view of past experiences:

> At the last meeting we were told, "There will be a 50/50 chance that we will get above-average rainfall and 50/50 chance we will get below-average rainfall." This told us nothing.

The degree to which these farmers incorporated seasonal forecasting information into their decision making was also variable. Although no one used it as a regular aid, some said they sometimes used it, others considered it but rarely used it, and some said they did not use it at all.

Those who said they did use it indicated that it could affect changes in planning for the timing of seeding, spraying times, planting rates, the type of crops planted, and the number of stock purchased. However, they stressed that the seasonal forecasting was only one piece of information they used, combining it with natural indicators and sources of information used in the past. Decisions were still very conservative.

> If they say it is going to be a dry year I won't buy more cattle. If it is going to be a wet year, I may decide to buy more cattle.

Those who did not use seasonal forecasting in their decision making were reluctant to do so because they had bad experiences in the past, or had heard of someone who had. Other reasons included a lack of understanding or restricted access to the information. They seemed to pay more attention to short-term forecasting than to seasonal. The consequences of poor short-term decisions were not seen to be as dire as the consequences of seasonal mistakes.

> Really if your gut feeling tells you it is going to be dry then it probably will be. If it tells you it will be wet, it probably will be.

> I'm an old-time farmer and I feel that you take what you get.
>
> I don't have much confidence in the information. It is usually only 50% accurate which is the same as tossing a coin.

Rainfall probabilities were considered to be the most useful information seasonal forecasting could provide. However, decision making would still be highly conservative.

> I would pay attention if they told me there was a 75% chance that we will go into a drought. However, if they told me that there was a 25% chance of below-average rainfall, with a 75% chance of above-average rainfall, I would pay more attention to the prediction of below-average rainfall.

The farmers were asked if they would be more willing to use a tool that predicted only extreme events with better than usual reliability, rather than more frequent rainfall predictions with lesser certainty. Generally it was agreed that this would be preferable, but there was considerable cynicism that sufficient reliability could be obtained for their purposes.

They did acknowledge, however, the difficulty associated with forecasting, especially given the limited recorded weather history in Australia. Then again, it seemed there was little likelihood that any latitude would be given to the scientists if the forecasts were mistaken. Reliability was very important, and until this could be achieved to help farmers in decision making, uptake of the technology would be limited. And memories can be very long.

> Indigo Jones was a long-range forecaster a while back, and he was considered to be very good. In 1974 he predicted it would be wet and we had some of the biggest floods in history. However, in 1975 he predicted it would be wetter still, and we had one of the worst droughts on record. After that I lost faith in long-range forecasters.

It is therefore apparent that the potential market for seasonal forecasting tools in the farming community will be limited in the short term. One must understand the likely users of the technology and exactly what decisions they believe it can assist them with.

IV. SUMMARY

Regions with high interannual variability of streamflow present challenges for managing the risks associated with the use of their water resource systems. Reliable forecasts of streamflow months in advance offer opportunities to manage this risk. Knowledge of more or less streamflow than average has the potential to influence farmer decision making and/or the allocation of water to the environment or to replace groundwater stocks under threat. In the long term, it has the potential to improve the viability of agricultural production activities and increase water use efficiency while maintaining desirable environmental flows.

The ability of the integrated modeling approach described above to provide comparison with alternate heuristic forecasting techniques has demonstrated the practical advantage of forecasting methods over these alternate techniques. In the overwhelming number of cases, where water management and consequent planting decisions were based on seasonal forecasts, the enterprise would have returned a better result in terms of water use efficiency and net gains in profit. Yet adoption is slow, perhaps reflecting the general conservative nature of farmers, at least in Australia, and the need for them to see real and sustained benefit before they will consider incorporating such tools into their decision making.

The social analysis confirmed that knowledge and understanding of the term and the forms of seasonal forecasting were highly variable. There seemed to be a misunderstanding of the difference between types of forecasting and sources of forecasting. However, there was considerable support for natural signals rather than the use of technology. The degree to which people understood probabilities, or thought they could be useful, was also variable. Many were skeptical of the probabilities given their derivation by the extrapolation of past data to the present. Their usefulness was also questioned in view of past experiences.

An aggressive water reform agenda, underpinned by an acknowledgment of the finite size of the water resource and recognition of the legitimacy of the environment as a water user, is driving research in and development of tools that can fine-tune critical decisions about water allocation.

V. FUTURE DIRECTIONS

In spite of recent improvements, adoption of climate variability management tools such as seasonal forecasting is low among the farming community. Farmers have expressed negativity about the reliability of the tools and their benefits.

It could be argued that many farmers, particularly those in large irrigated enterprises, have already reduced the risk associated with timely access to water by building large on-farm water storages and installing more efficient water reticulation systems—that is, they have invested (at a significant cost) out of the uncertainty for which seasonal forecasting tools are trying to compensate. So, it would seem that the need to consider the use of technology such as seasonal forecasting is directly related to degree of exposure and risk management behavior.

A consequence of the current water reforms—as timely access to instream water is no longer guaranteed and on-farm water storage is increasingly regulated—is an increase in risk exposure. This may force users to invest in tools that provide marginal gains. To identify where these marginal gains are, and the different levels of benefit that are possible, forecasting tools need to be tailored to a range of niche markets whose needs, decision-making behaviors, and current resistance must be clearly articulated. These niche markets need to be identified. This case study gives some strong leads—for example, irrigation versus dryland and high-equity versus mortgage participants. Significant benefit from research and development in climate risk management can only be realized if it produces tools that match users' needs and expectations and that can be incorporated into their decision-making and risk assessment processes.

ACKNOWLEDGMENTS

The opportunity to form a cohesive research team of leading Australian experts in the fields of hydrology, climatology, seasonal forecasting, integrated assessment, and social analysis was made possible via an 8-month research grant from the Managing Climate Variability Program within Land and Water Australia, the government agency responsible for land and water resources research and development in Australia. Such integration demonstrates the gains that can be made by adopting an interdisciplinary approach to developing practical tools to support sound environmental management.

REFERENCES

Chiew, FHS; McMahon, TA. Global ENSO-streamflow teleconnection, streamflow forecasting and interannual variability. *Hydrological Sciences Journal* 47:505–522, 2002.

Chiew, FHS; McMahon, TA. Australian rainfall and streamflow and El Niño/Southern Oscillation. *Australian Journal of Water Resources* 6:115–129, 2003.

Chiew, FHS; Peel, MC; Western, AW. Application and testing of the simple rainfall–runoff model SIMHYD. In: VP Singh, DK Frevert, eds. *Mathematical Models of Small Watershed Hydrology and Applications* (pp. 335–367). Littleton, CO: Water Resources Publication, 2002.

Clewett, JF; Clarkson, NM; George, DA; Ooi, S; Owens, DT; Partridge, IJ; Simpson, GB. *Rainman StreamFlow* (version 4.3): A Comprehensive Climate and Streamflow Analysis Package on CD to Assess Seasonal Forecasts and Manage Climatic Risk. QI03040.Department of Primary Industries, Queensland, 2003.

Letcher, RA. Issues in integrated assessment and modeling for catchment management. Ph.D. thesis. The Australian National University, Canberra, 2002.

Nash, JE; Sutcliffe, JV. River forecasting using conceptual models, 1. A discussion of principles. *Journal of Hydrology* 10:282–290, 1970.

Peel, MC; McMahon, TA; Finlayson, BL; Watson, FGR. Identification and explanation of continental differences in the variability of annual runoff. *Journal of Hydrology* 250:224–240, 2001.

Piechota, TC; Chiew, FHS; Dracup, JA; McMahon, TA. Development of an exceedance probability streamflow forecast using the El Niño–Southern Oscillation and sea surface temperatures. *ASCE Journal of Hydrologic Engineering* 6:20–28, 2001.

Sharma, A. Seasonal to interannual rainfall probabilistic forecasts for improved water supply management: Part 3—A nonparametric probabilistic forecast model. *Journal of Hydrology* 239:249–259, 2000.

URS Australia. Defining researching opportunities for improved applications of seasonal forecasting in south-eastern Australia with particular reference to the southern NSW and Victorian grain regions. Report to the Climate Variability in Agriculture Program, Land and Water Australia, Canberra, 2001.

14

Droughts and Water Stress Situations in Spain

MANUEL MENÉNDEZ PRIETO

CONTENTS

I. MAIN CHARACTERISTICS OF DROUGHTS IN SPAIN

A. Introduction: Water Management and Planning Framework in Spain

In Spain, water is not just another natural resource. Drought is more than a combination of meteorological factors because it usually produces conflicts between users, deterioration in river ecology, and increased public awareness. During drought, the decreased availability of water results in greater pressure on existing surface and subsurface water supplies, and debates about potential remediation measures usually go beyond the scientific or technical spheres and into the political sphere.

Water management in Spain is based on river basin departments (RBDs) (see location of main Spanish river basins in Figure 1). If water runs through more than one regional territory (autonomous community), the RBD depends on the state administration and is called an *inter-community*

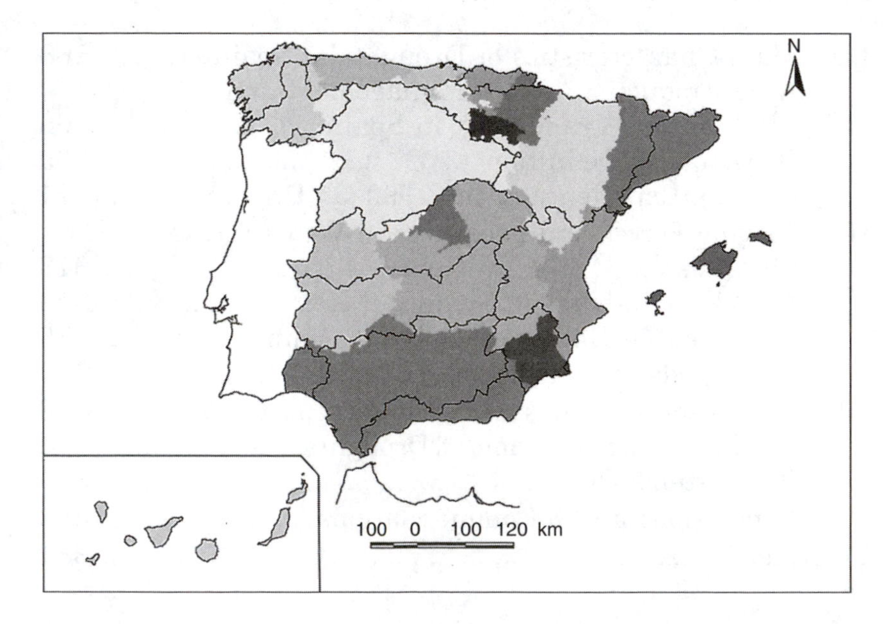

Figure 1 Location of main river basin departments (black border lines) and autonomous communities (shaded regions).

basin. An *intra-community basin* lies completely within the limits of one autonomous community, and the regional administration assumes authority for water management.

Water planning is accomplished through two main instruments: basin hydrological plans and the national hydrological plan.

The Water Act of 1985 conceived the basin hydrological plans as the central instruments in regulating water, to which, according to the text of the act, "all action in the public domain is subject." After a drafting process of more than 10 years, the legal texts of the plans were finally approved in 1998. Afterward, information about water resources in Spain was collected, analyzed, and described in the so-called "Libro Blanco del Agua en España" ("White Paper on Water in Spain") (Ministerio de Medio Ambiente, 2000), and it was used as the basis of the national hydrological plan. Approved by the Spanish Parliament in July 2001 and partially modified in 2004, the national hydrological plan includes the following measures to ensure the coordination of the different basin hydrological plans: forecasts, structural measures, and modifications affecting existing public supply or irrigation systems.

B. Drought Definition

As in other Mediterranean countries with large arid or semiarid areas, droughts in Spain are difficult to evaluate and quantify and thus are difficult to define. Many definitions are used, and often it is not clear when a drought situation has started or finished or even if it has existed. In some large river basins, the definition of a drought is based on simple rainfall statistics. For example, for the river Ebro, one of the largest rivers in Spain, a dry period starts according to Spanish law, "when rainfall amounts in two consecutive months within the series are lower than 60% of the average rainfalls for these months." For the river Guadiana, "a situation of drought occurs when the sum of rainfalls registered during the 12 preceding months is lower than those registered in 75% of the cases within the period analyzed." In some cases, the definition of drought is based on the relationship between supply and demand. For instance, in the Guadalquivir River basin, the following definition is used: "a situation in which the resources accumulated are insufficient to satisfy the demand." In the river basin Norte III* (one of the wettest in Spain), the experience of previous droughts must be taken into account to plan water supply facilities. The Norte III hydrological plan points out "the works have to be planned in accordance with the demand on the basis of the droughts that occurred in the 1941–43 and 1989–90 periods, without admitting any errors and considering that there are sufficient resources for such purposes within the scope of the plan." This plan also defines "dry year" as one in which the yearly cumulative flow is half of the average and "very dry year" as one in which the cumulative flow amounts to 75% of the flow registered in a dry year, or slightly more than 35% of the yearly average cumulative flow.

Regarding drought definition, an interesting case refers to the provisions about emergency situations included in the treaty between Spain and Portugal on transboundary river basins. This agreement, known as Albufeira, was signed in Albufeira, Portugal, at the end of 1998. It establishes that in case of drought situations, Spain will not be in breach of the treaty for failing to maintain the quantity or quality of the discharges fixed in the agreement as minimum inflow to Portugal for *standard meteorological conditions*. In a drought situation, it is understood that the first priority should be to maintain the water supply for urban uses and, in consequence, other requirements of the treaty would be temporarily repealed. Drought situations are defined on the basis of a referenced precipitation calculated for a specific time period in each river basin using data from only two or three rain gauges through different weights that are applied at each control point. This precipitation is compared with a percentile of the mean for the same period. Tables 1 and 2 explain the drought definition process proposed for the river Miño.

---

* Norte river basin is divided into three subbasins for administrative purposes: Norte I, Norte II, and Norte III.

Table 1 Step 1: Calculation of the Mean (M)[a] and Reference (R) Precipitation

River Basin	Rain Gauge Stations	Weights of Each Control Point for the Precipitation Assessment
Miño	Lugo	30%
	Orense	47%
	Ponferrada	23%

[a] Calculated for the period 1945–46 to 1996–97 and updated every 5 years.

Table 2 Step 2: Identification of a Drought Situation

River Basin	Control Point (along border between Spain and Portugal)	Minimum Discharge to Be Ensured by Spain along the Border with Portugal (millions m³/year)	Starting Date for Derogation Period is July 1 if …	Ending Date for Derogation Period is the Following Month to December if …
Miño	Salto de Freira	3700	R (cumulative rainfall from October 1 to July 1) <70% M (cumulative rainfall from October 1 to July 1)	R (cumulative rainfall from October 1 to July 1) >M (cumulative rainfall from October 1 to July 1)

The lack of a universally accepted definition for drought has consequences that go beyond academic discussions. For example, as will be indicated in the following section, Spanish water regulations allow river authorities to adopt emergency measures, including occupation of private lands or stopping specific uses such as irrigation, if a drought situation has been declared.

C. The Experience of the 1990–95 Drought

The most severe droughts in Spain in the last century occurred in 1941–45, 1979–83, and 1990–95. These three droughts were extensive and affected most of the country. The map in Figure 2 shows the percentage decrease in average rainfall from October 1990 to September 1995 compared to the average rainfall for the 1940–1996 period. Several river basins suffered decreased rainfall percentages, around 30%. The resulting reduction in runoff in most of the country was more than 40% and amounted to more than 70% in two basins (Guadiana and Guadalquivir).

The drought of 1990–1995 affected more than one-third of the total population of Spain. For example, during these years, there were severe restrictions on water use in cities like Granada, Malaga, and Seville, and irrigation was banned during 1993–95 in the Guadalquivir River basin. The situation forced cities and regions to adopt different measures. For example, the search for new groundwater resources was car-

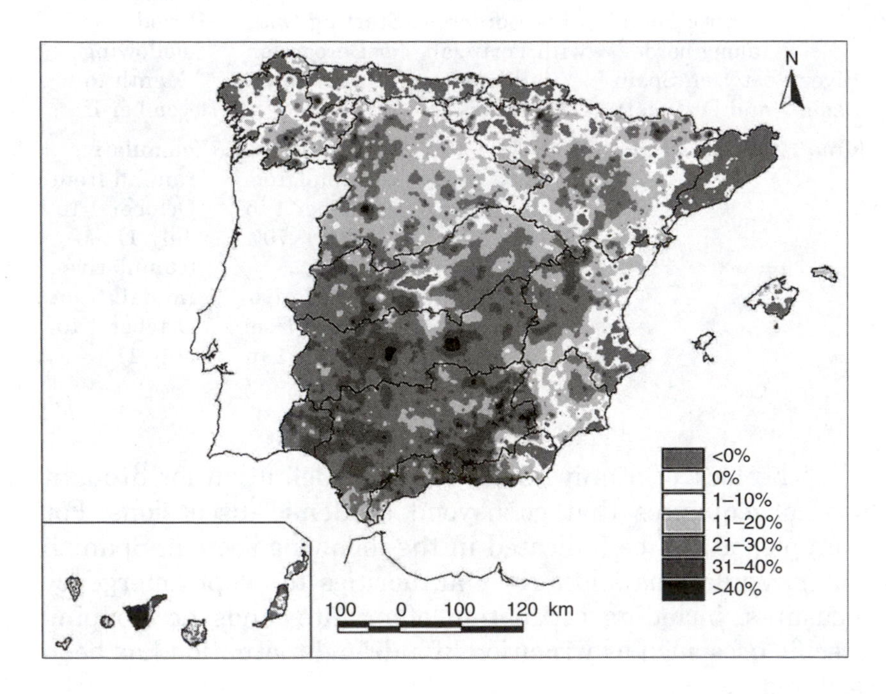

Figure 2 Percentage decrease in rainfall during 1990–1995 in Spain in comparison with the average values for the period 1940–1996.

ried out on a large scale. The city of Granada, with a population of 300,000 and a yearly demand of around 35 million m^3, had to be supplied mainly with groundwater, changing its water supply system, which had previously used mainly surface water. Groundwater was also used to solve problems in irrigated areas such as the Jucar and Segura River basins. Unconventional methods were applied, such as increasing water reuse and even transporting water by boat to cities like Cadiz and Majorca. New desalinization plans were proposed, but most were never carried out. Regulations were adopted to allow emergency measures to be taken. For example, the so-called Commissions on Droughts in each RBD had the power to reduce or suspend water use and require users to install specific devices for saving water. A decree issued by the Council of Ministers empowered the government, after consulting the Commissions on Droughts, to adopt "such measures as shall be necessary as regards the utilization of the public water domain, even if a concession has been granted for its exploitation," if an extraordinary drought occurs.

II. DRIVING FORCES AND PRESSURES IN WATER STRESS SITUATIONS

A. Water Balance in Spain

Spain on average has sufficient water to meet water demands, but these resources are very unevenly distributed. According to the "White Paper on Water in Spain" (Ministerio de Medio Ambiente, 2000), natural water resources are estimated at 111 billion m^3 per year, distributed as shown in Figure 3.

The total amount of natural water resources is not available for abstractions.* Two limitations are usually considered: (1) a precautionary reserve of 20% of the natural resources for environmental requirements, and (2) the minimum water volume that must be maintained along the Spanish–Portuguese border, per the Albufeira Treaty.

Annual freshwater abstraction by sector is shown in Table 3 (Ministerio de Medio Ambiente, 2000). To illustrate the possibilities of reusing water, we obtain the consumptive

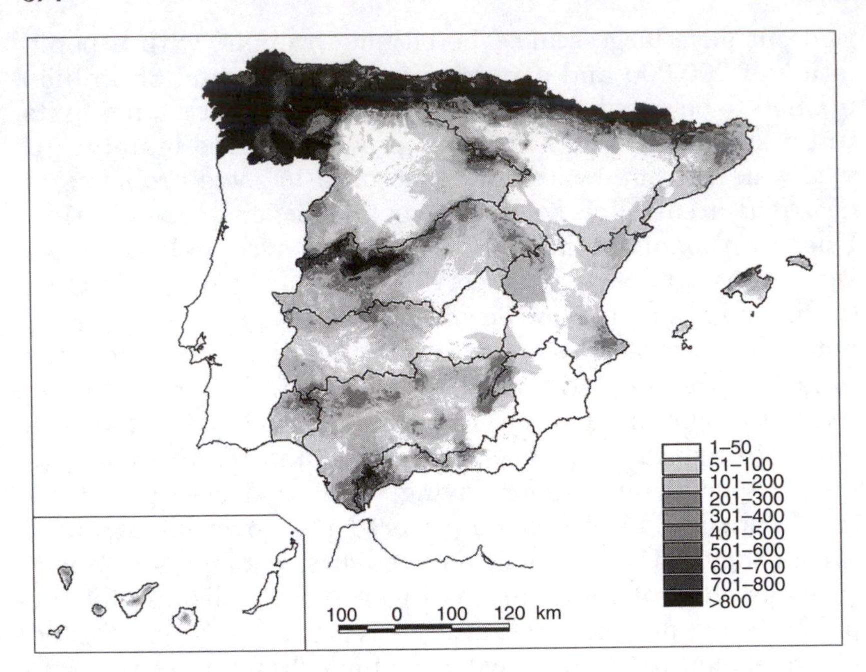

Figure 3 Natural annual water resources in Spain (mm). (From Ministerio de Medio Ambiente, 2000.)

fraction of the water demand that cannot be reused, by applying different percentages: 20% for urban use, 80% for irrigation, 20% for industrial use, and 5% for energy use.

With the percentages for reusing water and data obtained from the basin hydrological plans, the distribution of the total water demand in Spain is shown in Figure 4.

Once maps of potential water resource supply and demand have been drawn up, they can be compared to identify existing imbalance and its territorial location (see Figure 5).

---

* EUROSTAT (Statistical Office of the European Communities) defines *water abstraction* as "water removed from any source, either permanently or temporarily." In Europe, *water abstraction* is equivalent to *water withdrawal*. *Water demand* is defined as the "volume of water requested by users to satisfy their needs." According to EUROSTAT, *water use* is "water actually used by end users for a specific purpose within a territory, such as for domestic use, irrigation or industrial processing."

TABLE 3 Annual Freshwater Abstraction in Spain, by Sector

Urban Water Demand	Consumptive Urban Water Demand	Agricultural Water Demand (Irrigation)	Consumptive Agricultural Water Demand (Irrigation)	Industry Water Demand*
5,393	1,079	27,863	22,290	1,920

Consumptive Industrial Water Demand	Energy Water Demand	Consumptive Energy Water Demand	Total Water Demand	Total Consumptive Water Demand	Returned Water
384	5,679	284	40,855	24,037	16,818

Note: All figures are in millions of cubic meters per year.
Source: Ministerio de Medio Ambiente (2000).

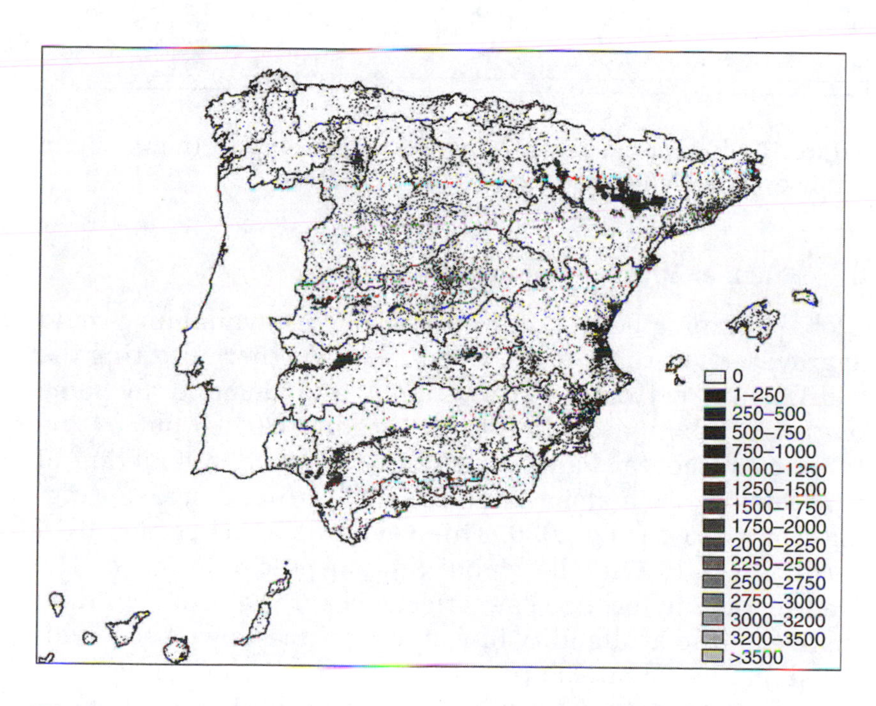

Figure 4 Total water demand in Spain (mm). (From Ministerio de Medio Ambiente, 2000.)

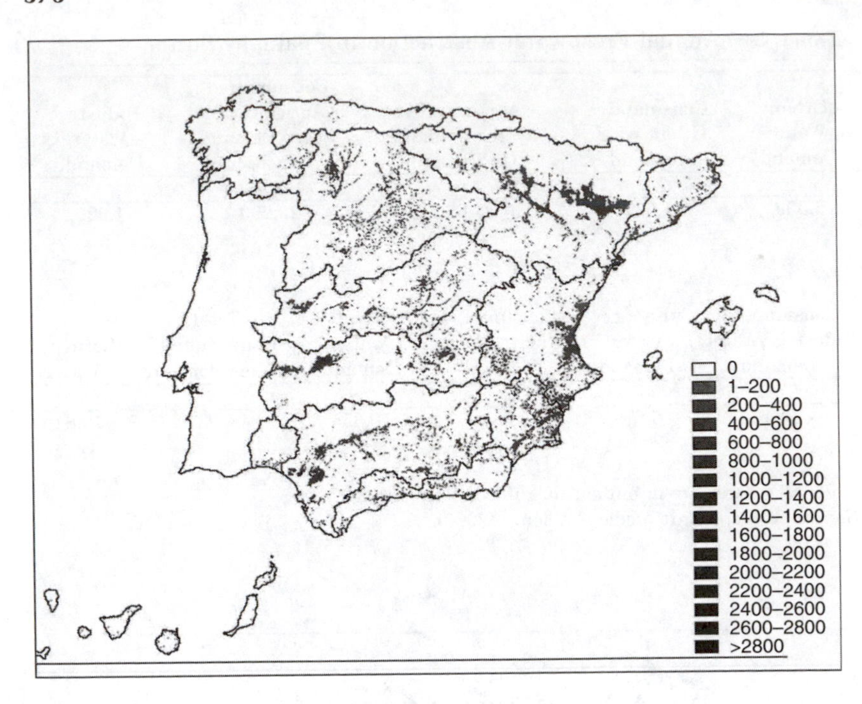

Figure 5 Territorial distribution of water deficits (mm). (From Ministerio de Medio Ambiente, 2000.)

B. Potential Effects of Global Warming

Global warming effects on water resources availability could aggravate the situation described above, further upsetting the equilibrium between water supply and demand in some places. In Spain, according to the studies carried out by the Intergovernmental Panel on Climate Change, a doubling of CO_2 could cause a global annual temperature increase ranging from 1 to 4°C by 2030. This would be accompanied by a decline of 5–15% in the global annual precipitation for the Mediterranean basin. The effects resulting from climate change on the availability of water resources have been evaluated in the "White Paper on Water in Spain" under two scenarios. An optimistic scenario, Scenario 1, indicates an increase of 1°C in the mean annual temperature. Scenario 2, a more pessimistic scenario, indicates a decline of 5% in the average annual precipitation and an increase of 1°C in temperature. Scenario 1 produces a decline of about 5% in water resources for the entire territory of Spain, with reductions around 10% in rivers such as Guadiana, Júcar, and Segura. Scenario 2 causes a decrease of about 15% in the total water resources, with reductions of about 20% in the three rivers mentioned above.

C. Trends in Water Demands

Lack of water availability can be produced by a lower amount of resources in the system or by an increase in the demand. Higher water demands introduce new uncertainty factors and make some areas particularly vulnerable to drought situations. In many places, overexploitation of resources exacerbates drought impacts.

As in other Mediterranean countries, many areas of Spain do not have adequate water resources to meet all demands. In case of water conflicts in Spain, the Water Act of 1995 describes a water use list, establishing the following priorities, from first to last in importance: water supply in urban areas, irrigation, industrial uses for power generation, other industrial uses, fish farming, recreational uses, and navigation.

In Spain, irrigation exerts the greatest pressure on water resources. The quantity of water required for irrigation constitutes about 70% of the total demand in Spain, or about 8000 m^3/ha/year. Irrigation is a key economic activity in Spain, covering only 15% of the total agricultural land but accounting for more than half of the total agricultural production and one-third of the total salaries in the sector. In some regions, the agri-food sector generates more than 15% of the industrial employment. Irrigation and, in general, agricultural activities in Spain depend mainly on the so-called Common Agricultural Policy (CAP) of the European Union. The CAP is basically a policy to regulate markets of agricultural products inside the European Union. Initially its philosophy was based on setting a minimum guarantee price for products, which assured farmers some income for their crops, beyond price fluctuations. The failures of this policy, among them the generation of considerable surpluses in some agricultural products, led to strong criticism that has culminated in several reforms. The new CAP is based on a reduction of the institutional prices, establishment of compensatory aids to the producer (instead of the protection of the product via prices), and the establishment of environmental protection measures: prevention of diffuse pollution, afforestation of arable lands, etc.

The second-highest consumptive water demand in Spain is urban water supply, currently at about 5 billion m^3, which could be satisfied by available water resources. However, the supply system is not always reliable. Recent droughts in Spain, especially the one in 1990–95, have shown that in significant areas of Spain, the supply systems have failed, and therefore the supply guarantee, which in theory should be close to 100%, has been far from its goal. In fact, serious shortages persist in Spain, often caused by problems in the supply infrastructures. In some cases, facilities do not have enough capacity, especially in tourist areas with high seasonal populations. Additionally, the water delivery infrastructure, especially in small cities, is outdated and results in significant leakage.

Figure 6 shows the expected trend of water demand in Spain, by sector. Water demands are expected to change in small percentages in the future. Urban water supply is an illustrative example. Although the population will increase over the next few years, improved water use efficiency and water-saving practices produce only a slight decrease in demand as a final result.

III. RESPONSE STRATEGIES

A. Early Identification of Droughts

In Spain and other northern Mediterranean countries, drought forecasting based on climatic indicators is not operational. One of the most suitable indicators is the North Atlantic Oscillation (NAO), which shows the pressure gradient in the region. In the winter, when the NAO is low, westerlies are weaker and do not penetrate as far into Europe, so temperatures are influenced by cold high pressure located over Eurasia and precipitation is reduced (EEA, 2001

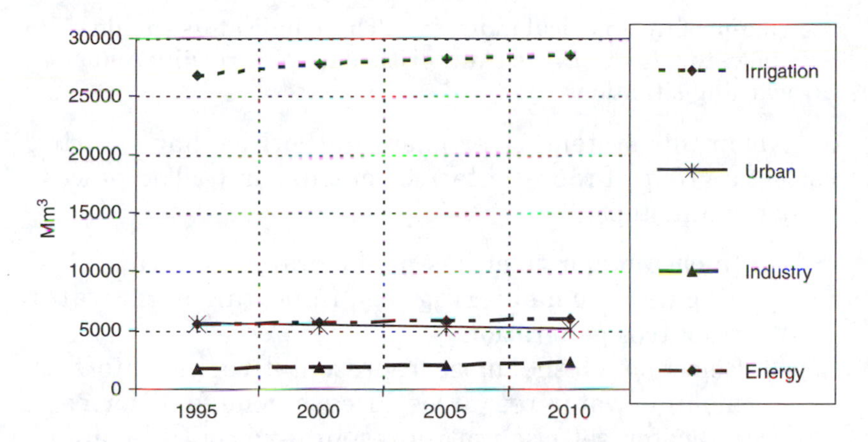

Figure 6 Trends in water demand in Spain. (From EEA, 1999.)

Forecasting the evolution of NAO is not yet possible. Some researchers (Iglesias, 2000) believe that reliable predictions in the Mediterranean area are far from being available for operational purposes.

Experience from recent severe droughts in Spain has shown that droughts are identified too late, complicating the implementation of remediation measures. The "White Paper on Water in Spain" (Ministerio de Medio Ambiente, 2000) notes:

> Taking into account the experience of recent droughts, it is advisable to implement an early warning system that can activate emergency plans. This early warning system should use indicators based on information easily available (precipitation, water stocks in reservoirs, or aquifer water table levels, for example) to alert people of the possible start date of a drought or to identify its intensity.

This approach has been adopted in the national hydrological plan of Spain. Title II (Complementary Rules for Planning) and article 27 (Drought Management) of the plan note:

> The Environment Department and river basins affecting two or more Autonomous Communities, looking for the reduction of environmental, economic, and social impacts produced by drought situations, will implement a global

system of hydrological indicators. These indicators should allow the River Basin Authorities to declare different warning situations.

To develop this system, river basin authorities have started to select a group of representative control points. The process has five main steps:

1. In each major river basin, elaboration of a zone classification, considering its importance in water resources generation
2. Selection of the most representative indicators to monitor water resources in each zone, and for each indicator, selection of representative control points
3. Collection of temporal data series for each control point
4. Elaboration of criteria to be used in early identification of droughts, according values to each indicator
5. Updating of the temporal series associated with each indicator (monthly data)

Preliminary classification of zones and selection of control points has been completed. The selection of control points includes:

- Rain gauge stations
- Streamflow gauging stations
- Reservoir stocks
- Water table indicators

Maps comparing the situation at the control points for the current month with historical data—for instance, by obtaining percentiles and indicating them by a color code—are the main output of the system. These maps, like the one shown in Figure 7, provide a general overview of the hydrological situation and can be used for defining drought severity or triggering emergency measures.

B. Systems Operation

The indicators previously described are mainly based on analysis of precipitation and flows. These two types of indicators

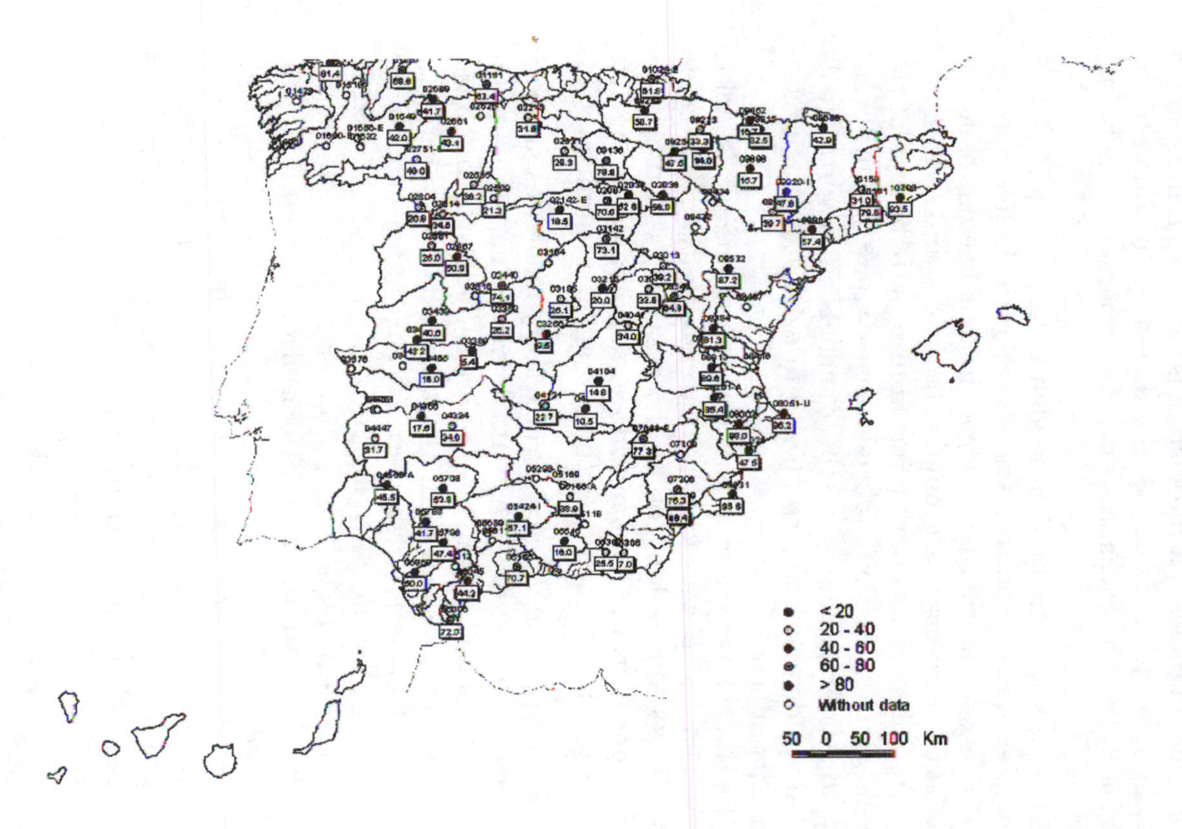

Figure 7 Selection of rain gauges for triggering emergency measures. (Example is for 1991.)

take into account only the water supply, not the demand.[*] For practical purposes, an indicator for a drought early warning system should indicate if the system is close to a failure, given that the main goal of the system should be to simulate the effect of mitigation measures or to put emergency plans into practice.

The use of mathematical modeling for the management of complex water resources systems (Sánchez Quispe et al., 2001) is a real improvement. These models use data such as water stock in reservoirs, groundwater levels, and streamflow. Their application for quantifying failure probabilities of the system elements allows a risk assessment for different demand scenarios. Therefore, these models offer worthy tools not only to detect drought but also to evaluate measures to reduce its effects.

In Madrid, the water supply company Canal de Isabel II has defined four phases for water stress situations (Canal de Isabel II, 2003) based on the total water stocks in the reservoirs managed by the company (see Figure 8) and the impacts of the water yield reduction on the management of the system.

Phase 0 or drought alert. Preparatory measures are triggered during this stage. Its definition ensures a minimum period of 2 months before the subsequent phase, even in the event of the lowest yields ever recorded.

Phase 1 or severe water stress. This stage defines the actual starting point for the drought. The main objective of this phase will be to reach a total reduction of 9% in consumptive uses, based on information campaigns.

Phase 2 or serious water stress. In this phase, several restrictions are imposed to reach an average reduction of up to 26% in consumptive uses (on average; values depend on the different water uses).

---

[*] However, reservoir stocks and water table indicators integrate supply and demand.

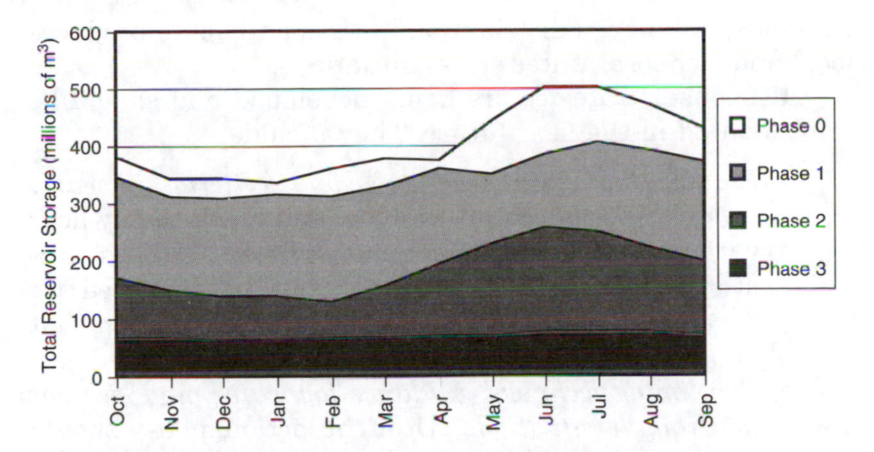

Figure 8 Water stress situation phases for the water supply company in Madrid for 2003. The figures are based on statistical analysis, so they must be updated yearly.

Phase 3 or emergency situation. This theoretical scenario would be a catastrophic social situation that would force the national government to apply civil protection measures. Water consumption should be reduced to 80 liters per capita per day.

IV. CONCLUSIONS AND RECOMMENDATIONS

Spain, like many other Mediterranean countries, has not had clear criteria and strategies to identify and manage droughts. In the severe drought of 1990–95 the crisis situation was identified too late, affecting the efficiency of emergency measures. In addition, drought forecasting based on climatic indicators is not operational at present.

A major increase in water demands in Spain is not foreseen because improved water use efficiency neutralizes the effect of population or irrigation growth. However, the effects of global warming on water resources availability could break the equilibrium between water supply and demands. Moreover, although Spain on average has sufficient water to meet demands, its uneven distribution in time and space produces local and temporal water stress situations.

Response strategies are being developed and should be implemented in the near future. They include:

- *Elaborating clear and consistent criteria for early drought identification*, which would allow timely activation of established emergency plans in a crisis situation. This early identification implies the need for developing alert indicators based on information that is usually available.
- *Developing technical guidance for water management in drought situations*. Drought action plans should have clearly established rules and procedures, including criteria for applying restrictions, conditions for adopting special procedures to increase the flexibility and exchange of rights between users and their financial regulation, mobilization of hydrologic reserve areas, and temporary increases in the exploitation of aquifers.

Finally, an extremely important drought preparedness issue is ensuring the transparency of the drought assessment process and continuous public participation in the decision-making process. The cooperation of end users in efforts to save water and design measures to be taken in emergency situations is essential.

REFERENCES

Canal de Isabel II. Manual del Abastecimiento del Canal de Isabel II, 2003.

EEA. Sustainable Water Use in Europe. Part 1. Sectoral use of water. Environmental issue report No. 1. European Environment Agency, 1999.

EEA. Sustainable Water Use in Europe. Part 3. Extreme hydrological events: Floods and droughts. Environmental issue report No. 21. European Environment Agency, 2001.

Iglesias A. Impacto del cambio climático en la agricultura: escenario para la producción de cultivos en España. *El Campo*, 137:183–201, BBVA, 2000.

Ministerio de Medio Ambiente. Libro Blanco del Agua en España, 2000.

Sánchez Quispe ST, J Andreu, A Solera. Gestión de recursos hídricos con decisiones basadas en estimación del riesgo. Universidad Politécnica de Valencia, 2001.

Part IV

Integration and Conclusions

15

Drought and Water Crises: Lessons Learned and the Road Ahead

DONALD A. WILHITE AND
ROGER S. PULWARTY

CONTENTS

I. INTRODUCTION

Despite the fact that drought is an inevitable feature of climate for nearly all climatic regimes, progress on drought preparedness has been extremely slow. Many nations now feel a growing sense of urgency to move forward with a more proactive, risk-based drought management approach (ISDR, 2003; Wilhite, 2000). Certainly the widespread occurrence of this insidious natural hazard in recent years has contributed to the sense of urgency. But, drought occurs in many parts of the world and affects portions of many countries on an annual basis. For example, the average area affected by severe and extreme drought in the United States each year is 14%. This figure has been as high as 65% (1934) and has hovered in the 35–40% range in recent years. So, does the widespread occurrence of drought in the United States over the last 5–6 years explain the emergence of several national initiatives centered on drought monitoring and preparedness, given that events of this magnitude have not motivated policy makers to act in the past? Our experience would suggest that this is only one of the factors contributing to the increased attention being directed to this subject in the United States and in other drought-prone countries.

Climate change and the potential threat of an increase in frequency and severity of extreme events are also a contributing factor. However, the uncertainty associated with climate change is probably not playing a significant role in this trend because most policy makers have difficulty thinking beyond their term of office or the next election. For regions that have in recent decades experienced either a downward trend of annual precipitation or a higher frequency of drought events (perhaps multi-year in length), or both, the potential threat of climate change already seems real. Decision making under uncertainty is onerous, but critically important. Policy makers and resource managers often seem of the opinion that climate change projections are in error, preferring to presume that there will not be a change in the climate state and that extreme climatic events such as drought will not change in frequency or severity. Little consideration is given to the real posibility fact that projected changes in climate may be too conservative or underestimate the degree of change in the frequency and severity of extreme events for some locations.

In our view, the most significant factor explaining the growing interest in drought preparedness is associated with the documented increase in social and economic vulnerability as exemplified by the increase in the magnitude and complexity of impacts. Although global figures for the trends in economic losses associated with drought do not exist, a recent report from the U.N. Development Program (UNDP Bureau of Crisis Prevention and Recovery, 2004) indicates that annual losses associated with natural disasters increased from US$75.5 billion in the 1960s to nearly US$660 billion in the 1990s. Losses resulting from drought likely follow a similar trend. In addition, these figures for natural disasters are likely underestimated because of the inexact reporting or insufficiency of the data. And, these loss estimates do not include social and environmental costs associated with natural disasters over time. This increase has been observed in both developing and developed countries, although the types of impacts differ markedly in most cases, as illustrated by numerous authors in this book.

With respect to drought, how can we define vulnerability? It is usually expressed in terms of a society's capacity to anticipate, cope with, resist or adapt to, and recover from the impact of a natural hazard. Vulnerability is represented by a continuum from low to high and varies among community, population group, region, state, and nation. It is the result of many social factors. For example, population is not only increasing but also shifting from humid to more arid climates in some areas and from rural to urban settings for most locations. As population increases and lifestyles change, so do the pressures on water and other natural resources. Conflicts between water users escalate accordingly. Population increases force more people to reside in climatically marginal areas where exposure to drought is higher and the capacity to recover is diminished. Urbanization is placing more pressure on limited water supplies and overwhelming the capacity of water supply systems to deliver that water to users, especially during periods of peak demand. An increasingly urbanized population is also increasing conflict between agricultural and urban water users, a trend that will only be exacerbated in the future. More sophisticated technology decreases our vulnerability to drought in some instances while increasing it in others. Greater awareness of our environment and the need to preserve and restore environmental quality is placing increased pressure on all of us to be better stewards of our physical and biological resources. Environmental degradation such as desertification is reducing the biological productivity of many landscapes and increasing vulnerability to drought events. All of these factors emphasize that our vulnerability to drought is dynamic and must be evaluated periodically. The recurrence of drought today of equal or similar magnitude to one experienced several decades ago will likely result in far greater economic, social, and environmental losses and conflicts between water users.

II. MOVING FROM CRISIS TO RISK MANAGEMENT: CHANGING THE PARADIGM

In 1986, an international symposium and workshop was organized at the University of Nebraska that focused on the principal aspects of drought, ranging from prediction, early warning, and impact assessment to response, planning, and policy. The goal of this meeting was to review and assess our current knowledge of drought and determine research and information needs to improve national and international capacity to cope with drought (Wilhite and Easterling, 1987). Reflecting on this meeting today, nearly 20 years later, and its outcomes, it would seem that it may represent the beginning of the movement to a new paradigm in drought management—one focusing on reducing societal vulnerability to drought through a more proactive approach. Today, we are experiencing the impacts of drought in greater magnitude than ever before. Clearly, it has taken time for the policy community to become more aware of these impacts, their complexities, and the ineffectiveness of the reactive, post-impact or crisis management approach. The factors explaining the slow emergence of this new paradigm are many, but it is clear that it has emerged in many countries and in many international organizations dealing with disaster management and development issues. A more proactive, risk-based approach to drought management must rely on a strong science component. It also must occur at the interstices of science and policy—a particularly uncomfortable place for many scientists.

Figure 1 in Chapter 6 illustrates the cycle of disaster management, depicting the interconnectedness or linkages between crisis and risk management. The traditional crisis management approach has been largely ineffective, and there are many examples of how this approach has increased vulnerability to drought because of individuals' (i.e., disaster victims') greater reliance on the emergency response programs of government and donor organizations. Drought relief or assistance, for example, often rewards the poor resource manager who has not planned for drought whereas the better resource manager who has employed appropriate mitigation tools is not eligible for this assistance. Thus, drought relief is often a disincentive for improved resource management. Should government reward good stewardship of natural resources and planning or unsustainable resource management? Unfortunately, most nations have been following the latter approach for decades because of the pressures associated with crises and the lack of preparation. Thus, relief can also mask fundamental underlying problems of governance and international policy. Redirecting this institutional inertia to a new paradigm offers considerable challenges for the science and policy communities.

As has been underscored many times by the contributors to this volume, reducing future drought risk requires a more proactive approach, one that emphasizes preparedness planning and the development of appropriate mitigation actions and programs, including improved drought monitoring and early warning. However, this approach has to be multi-thematic and multi-sectoral because of the complexities of associated impacts and their interlinkages. Risk management favorably complements the crisis management part of the disaster management cycle such that in time one would expect the magnitude of impacts (whether economic, social, or environmental) to diminish. However, the natural tendency has been for society to revert to a position of apathy once the threat accompanying a disaster subsides (i.e., the proverbial "hydro-illogical cycle"; see Figure 1 in Chapter 5).

This raises an important point that has not been addressed in detail in this publication: What constitutes a crisis? Crises are inextricably tied to decision making. The Merriam-Webster dictionary gives the following definitions of crisis: the decisive moment (as in a literary play); an unstable or crucial time or state of affairs whose outcome will make a decisive difference for better or for worse. The word crisis is taken from the Greek "krisis," which literally means "decision." A crisis may be said to be occurring if a change or cumulative impacts of changes in the external or internal environment generates a threat to basic values or desired outcomes, there is a high probability of involvement in conflict (legal, military, or otherwise), and there is awareness of a finite time for response to the external value threat. A crisis is not yet a catastrophe; it is a turning point. Crisis situations can be ameliorated if different levels of decision makers perceive critical conditions to exist and if a change of the situation is possible for the actors. Thus informed "decision making" is key to effective mitigation of crises conditions and the proactive reduction of risk to acceptable levels. Being proactive about hazard management brings into play the need for decision support tools to inform vulnerability reduction strategies, including improved capacity to use information about impending events.

A key decision support tool for crisis mitigation is embedded within the concept of "early warning." As discussed in Chapter 1 and elsewhere in this volume, early warning systems must be made up of several integrated subsystems, including (1) a monitoring subsystem; (2) a risk information subsystem; (3) a preparedness subsystem; and (4) a communication subsystem. Early warning systems are more than scientific and technical instruments for forecasting hazards and issuing alerts. They should be understood as credible and accessible information systems designed to facilitate decision making in the context of disaster management agencies (formal and informal) in a way that empowers vulnerable sectors and social groups to mitigate potential losses and damages from impending hazard events (Maskrey, 1997).

Natural hazard risk information, let alone vulnerability reduction strategies, is rarely if ever considered in development and economic policy making. Crisis scenarios can let us view risk reduction as much from the window of opportunity provided by acting before disaster happens as from the other smaller, darker pane window following a disaster. Given the slow onset and persistent nature of drought, mitigating potential impacts, in theory and in practice, must be recast as an integral part of development planning and implemented at national, regional, and local levels. Institutions responsible for responding to droughts must take a more proactive stance in assisting sectors through their own private and public institutions in preparing not only for disaster events but also in analyzing vulnerability and proposing practical pre-event mitigation actions. Impact assessment methodologies should reveal not only why vulnerability exists (who and what is at risk and why) but also the investments (economic and social) that, if chosen, will reduce vulnerability or risk to locally acceptable levels. Studies of the natural and social context of drought should include assessment of impediments to flows of knowledge and identify appropriate information entry points into policies and practices that would otherwise give rise to crisis situations (Pulwarty, 2003).

III. FINAL THOUGHTS

Drought results in widespread and complex impacts on society. Numerous factors influence drought vulnerability. As our population increases and becomes more urbanized, there are growing pressures on water and natural resource managers and policy makers to minimize these impacts. This also places considerable pressure on the science community to provide better tools and credible and timely information to assist decision makers. The adaptive capacity of a community (defined here in the broadest terms) means little if available tools, data, and knowledge are not used effectively. In addition, issues of sustainable development, water scarcity, transboundary water conflicts, environmental degradation and protection, and climate change are contributing to the debate on water management. Drought certainly exacerbates all of these problems and has significant cumulative impacts across all of these areas beyond the period of its climatological occurrence. Improving drought preparedness and management is one of the key challenges for the future. The motivation for this book was to provide insights into these important issues and problems and, it is hoped, point toward some real and potential solutions. The contributors to this volume have addressed a wide range of science, technology, and management issues in theory and practice. Building awareness of the importance of improved drought management today and investing in preparedness planning, mitigation, improved monitoring and early warning systems, and better forecasts will pay vast dividends now and in the future.

Finding the financial resources to adopt risk-based drought preparedness plans and policies is always given as an impediment by policy and other decision makers. However, the solution is right in front of them—divert resources from reactive response programs that do little, if anything, to reduce vulnerability to drought (and, as has been demonstrated, may increase vulnerability) to a more proactive, risk-based management approach. For example, in the United States, more than $48 billion has been spent on drought assistance programs since 1988 (David Goldenberg, personal communication). One can only imagine the advances that could have been made in pre-drought mitigation strategies had a substantial portion of these funds been invested in early warning and information delivery systems, decision support tools to improve decision making, improved seasonal climate forecasts, drought planning, impact assessment methodologies, and better monitoring networks. The key to invoking a new paradigm for drought management is educating the public—not only the recipients of drought assistance that have become accustomed to government interventions in times of crisis, but also the rest of the public, whose taxes are being used to compensate drought "victims" for their losses. There will always be a role for emergency response, whether for drought or some other natural hazard, but it needs to be used sparingly and only when it does not conflict with preestablished drought policies that reflect sustainable resource management practices.

On the occasion of World Water Day in March 2004, the Secretary-General of the United Nations, Kofi Annan, stated:

> Water-related disasters, including floods, droughts, hurricanes, typhoons, and tropical cyclones, inflict a terrible toll on human life and property, affecting millions of people and provoking crippling economic losses. ... However much we would wish to think of these as strictly natural disasters, human activities play a significant role in increasing risk and vulnerability. ... Modern society has distinct advantages over those civilizations of the past that suffered or even collapsed for reasons linked to water. We have great knowledge, and the capacity to disperse that knowledge to the remotest places on earth. We are also the beneficiaries of scientific leaps that have improved weather forecasting, agricultural practices, natural resources management, and disaster prevention, preparedness, and management. New technologies will continue to provide the backbone of our efforts. But only a rational and informed political, social and cultural response—and public participation in all stages of the disaster management cycle—can reduce disaster vulnerability, and ensure that hazards do not turn into unmanageable disasters.

We would argue that the complexities of drought and its differences from other natural hazards as outlined in this book are more difficult to invoke than for any other natural hazard, especially if the goal is to mitigate impacts. Special efforts must be made to address these differences as part of drought preparedness planning, or the differences will result in a failure of the mitigation and planning process. It is imperative that future drought management efforts consider the unique nature of drought, its natural and social dimensions, and the difficulties of developing effective early warning systems, reliable seasonal forecasts, accurate and timely impact assessment tools, comprehensive drought preparedness plans, effective mitigation and response actions, and drought policies that reinforce sustainable resource management objectives.

REFERENCES

Downs, A., Up and down with ecology: The "issue-attention" cycle, *The Public Interest*, 28, 38, 1972.

ISDR, Drought: Living with Risk (An Integrated Approach to Reducing Societal Vulnerability to Drought), ISDR Ad Hoc Drought Discussion Group, Geneva, Switzerland, 2003.

Maskrey, A., Report on National and Local Capabilities for Early Warning, International Decade for Natural Disaster Reduction Secretariat, Geneva, October 1997.

Merriam-Webster, New Collegiate Dictionary. G & C Merriam Company. Springfield, MA. 1977. p. 270.

Pulwarty, R.S., Climate and Water in the West: Science, Information and Decision making. *Water Resources Update* 124, 4–12, 2003.

UNDP Bureau of Crisis Prevention and Recovery, *Reducing Disaster Risk: A Challenge for Development (A Global Report)*. New York: UNDP, 2004.

Wilhite, D.A. (ed.), *Drought: A Global Assessment,* Volumes 1–2, Hazards and Disasters: A Series of Definitive Major Works, edited by A.Z. Keller. London: Routledge Publishers, 2000.

Wilhite, D.A. and W.E. Easterling (eds.), Planning for Drought: Toward a Reduction of Societal Vulnerability, Proceedings of the International Symposium and Workshop on Drought, Westview Press, Boulder, CO, September 1987.

Index

D

E

V

W

Z